T0391567

Applying the Rasch Model and Structural Equation Modeling to Higher Education

This book introduces the fundamentals of the technology satisfaction model (TSM), supporting readers in applying the Rasch model and structural equation modeling (SEM) – a multivariate technique – to higher education (HE) research. User satisfaction is traditionally measured along a single dimension. However, the TSM includes digital technologies for teaching, learning and research across three dimensions: computer efficacy, perceived ease of use and perceived usefulness. Establishing relationships among these factors is a challenge. Although commonly used in psychology to trace relationships, Rasch and SEM approaches are rarely used in educational technology or library and information science. This book, therefore, shows that combining these two analytical tools offers researchers better options for measurement and generalisation in HE research. This title presents theoretical and methodological insights of use to researchers in HE.

Applying the Rasch Model and Structural Equation Modeling to Higher Education
The Technology Satisfaction Model

A.Y.M. Atiquil Islam

CRC Press
Taylor & Francis Group
Boca Raton London New York

CRC Press is an imprint of the
Taylor & Francis Group, an **informa** business

A CHAPMAN & HALL BOOK

Designed cover image: Istock

First edition published 2023
by CRC Press
4 Park Square, Milton Park, Abingdon, Oxon, OX14 4RN

and by CRC Press
6000 Broken Sound Parkway NW, Suite 300, Boca Raton, FL 33487–2742

© 2023 A.Y.M. Atiquil Islam

CRC Press is an imprint of Informa UK Limited

The right of A.Y.M. Atiquil Islam to be identified as the author of this
work has been asserted in accordance with sections 77 and 78 of the
Copyright, Designs and Patents Act 1988.

All rights reserved. No part of this book may be reprinted or reproduced or
utilised in any form or by any electronic, mechanical, or other means, now
known or hereafter invented, including photocopying and recording, or in
any information storage or retrieval system, without permission in writing
from the publishers.

For permission to photocopy or use material electronically from this
work, access www.copyright.com or contact the Copyright Clearance
Center, Inc. (CCC), 222 Rosewood Drive, Danvers, MA 01923,
978–750–8400. For works that are not available on CCC please contact
mpkbookspermissions@tandf.co.uk

Trademark notice: Product or corporate names may be trademarks or
registered trademarks, and are used only for identification and explanation
without intent to infringe.

British Library Cataloguing-in-Publication Data
A catalogue record for this book is available from the British Library

ISBN: 978-1-032-47140-2 (hbk)
ISBN: 978-1-032-47141-9 (pbk)
ISBN: 978-1-003-38472-4 (ebk)

DOI: 10.1201/9781003384724

Typeset in Bembo
by Apex CoVantage, LLC

Contents

Author   vii

1   **Assessment of ICT in Higher Education Applying the TSM**   1

     1.1 Introduction 1
     1.2 Literature Review 3
        1.2.1 Hypotheses of the TSM 5
     1.3 Method 7
     1.4 Findings 14
     1.5 Discussion 20
     1.6 Conclusion 22
        1.6.1 Implications for Policy and Practice 22
     1.7 Acknowledgements 23
     1.8 References 23
     Appendix 28

2   **Testing Online Learning Satisfaction in Higher Education**   40

     2.1 Introduction 40
     2.2 Background 42
     2.3 Literature Review 43
     2.4 Hypotheses 44
     2.5 Methodology 46
     2.6 Results 48
     2.7 Discussion 59
     2.8 Conclusion 61
     2.9 Acknowledgements 62
     2.10 References 62
     Appendix 68

3   **Assessing Online Research Databases in Higher Education Using the TSM**   75

     3.1 Introduction 75
     3.2 Review of the Literature 76
        3.2.1 Research Framework 78
           3.2.1.1 Research Hypotheses 80

vi *Contents*

3.3 *Methodology 83*
3.4 *Results 85*
3.5 *Estimating the Structural Model 90*
3.6 *Discussion 91*
3.7 *Conclusion 94*
3.8 *Acknowledgements 96*
3.9 *References 96*
Appendix 101

**4   Measurement of Wireless Internet in Higher Education
Using the TSM**   105

4.1 *Introduction 105*
4.2 *Background 106*
4.3 *Literature Review 108*
4.4 *Methodology 111*
    4.4.1 *Instrument 111*
        4.4.1.1 *Instrument Reliability and Validity 112*
4.5 *Results 113*
    4.5.1 *Estimating the Technology Satisfaction Model 116*
        4.5.1.1 *The Revised Technology Satisfaction
                Model 116*
        4.5.1.2 *Cross-Validation of the Technology Satisfaction
                Model 119*
4.6 *Discussion 121*
4.7 *Conclusion 125*
4.8 *Acknowledgements 126*
4.9 *References 126*
Appendix 131

Index   138

Author

A.Y.M. Atiquil Islam is Associate Professor in the Department of Education Information Technology at East China Normal University. He earned a multidimensional PhD by combining two faculties – education and computer science and information technology – at the University of Malaysia. His career has seen international collaboration on conducting quantitative research in higher education (including the USA, UK, Malaysia, China, Saudi Arabia, Oman, the Philippines, Iraq, Algeria, Pakistan and Bangladesh).

1 Assessment of ICT in Higher Education Applying the TSM

1.1 Introduction

The procedures of identifying, gathering and interpreting information about learning outcomes are central to what "assessment" means in educational contexts. Therefore, measurement directed towards what is important in such processes is critical. In general, assessment is an indispensable aspect of teaching, training and learning. Moreover, it is widespread within the systematic support of learning, both in training and in formal education (Farrell & Rushby, 2016). According to Bennett et al. (2017), assessment is an important factor in student engagement, as it has a critical impact on both students' learning and its certification. Furthermore, it could be highly beneficial to consider assessment design as a procedure supporting formative development rather than multiple iterations. Doing this could make the initial stakes lower, achieve a gradual roll-out over time, predict possibilities for collecting evidence and facilitate the management of resourcing and workloads. Such strategies are familiar to large educational projects or instructional design work but, comparatively, are much less frequently used in routine design work.

Assessment studies have long been prominent in all aspects of educational psychology (Baars et al., 2018; Graham, 2018; Halberstadt et al., 2018; Lee & Vlack, 2018; Martin & Lazendic, 2018; Mok et al., 2017; Zhu et al., 2018). Its emphases have covered, but have not been limited to, improvement in the quality of teaching, learning, research and the pursuit of research productivity. Contemporary psychologists have conducted numerous assessment investigations that apply computer-based analytical tools such as structural equation modeling (Lee & Vlack, 2018; Gibbons et al., 2018; Scherer et al., 2017; Zhu et al., 2018), hierarchical linear modeling (Areepattamannil & Khine, 2017), the Rasch model (Mok et al., 2015; Campbell & Bond, 2017), partial least squares (Onn et al., 2018), and computerised adaptive testing (Huebner et al., 2018; Buuren & Eggen, 2017). However, despite a decade of practice, the literature suggests that researchers or psychometricians may need to better comprehend and apply these analytical tools to improve the quality of practice measurements in higher education. In particular, the combination of structural equation modeling (SEM) and the Rasch model has not been adequately pursued in modern assessment strategy. Moreover, assessment studies have rarely taken

DOI: 10.1201/9781003384724-1

place in the arena of information and communication technologies (ICTs) as effective resources for tertiary education. In the arena of educational technology, researchers frequently use exploratory factor analysis (EFA) and SEM to perform assessments that focus on the reliability and validity of items rather than on the individuals being studied. The Rasch model considers both the items and the respondents in developing and validating the instrument using dichotomous and polytomous data. In such a case, this study assumed that combining these two analytical tools would provide a great opportunity for conducting meaningful measurements, which could potentially be generalised to and used in other situations.

Coniam and Yan (2016) claimed that the widespread development of ICT has had progressively powerful effects on numerous aspects of education, including assessment. Hence, it is assumed here that Chinese university lecturers' ICT skills would be valuable for performing their measurement activities, if they frequently used ICT for teaching, research and learning purposes. Despite the ubiquity of ICT applications, there has been a scarcity of investigations into the contributing dimensions of lecturer satisfaction with ICT as encountered in the tertiary education environment (Islam, 2015). According to Wu et al. (2010), *satisfaction* is the most recognised measure of the quality and usefulness of teaching and learning. Along these lines, Islam et al. (2019) identified that teachers' satisfaction depends on the benefits of using digital technologies. For instance, the effect of teachers' perceived usefulness of and satisfaction with new technologies in teaching and research manifested in teachers increasing their research productivity, enhancing their research skills, making information easier to find and providing the latest information on specific areas of research.

In the Chinese educational setting, explorations related to the pedagogical use of technologies have grown exponentially (Teo et al., 2018). However, the majority of studies have considered pre-service (Teo et al., 2018) and in-service (Teo & Zhou, 2017) teachers' use of technology. On the other hand, considering ICT's underlying benefits, like promoting educational and academic achievement, China aims to speed up the digitalisation process in the field of education. China intends to do this through engaging educational digitalisation within a developmental strategy of national digitalisation conceived in a holistic perspective, and it is currently implementing an educational information network. Hence, it is necessary to determine what facets of the current situation are affecting Chinese educators' uptake of technologies for educational purposes (Teo et al., 2018). Although technology plays a crucial role in promoting efficient instruction, there is still evidence that teachers do not always apply technology in a way that maximises its effect on teaching and learning (Teo & Zhou, 2017). In such a case, satisfaction could be one of the reasons why teachers or lecturers are not willing to use ICT frequently in their teaching and research activities in China. Therefore, this study aims to assess the components of ICT satisfaction for Chinese lecturers' through the technology satisfaction model (TSM) by applying structural equation modeling and the Rasch model.

1.2 Literature Review

In the contemporary literature on the acceptance or adoption of ICT application, the most recognised approach is the technology acceptance model (TAM) claimed by Davis et al. (1989). Technology acceptance studies have been globally conducted using the TAM (Yim et al., 2019; Al-Azawei et al., 2017; Fathali & Okada, 2018; Sánchez-Prieto et al., 2017; Holzinger et al., 2011). The TAM was generated from the theory of reasoned action (TRA) as first presented by Fishbein and Ajzen (1975), which is shown in Figure 1.1.

The TAM was designed to measure the causal associations among six dimensions such as external variables, perceived usefulness, perceived ease of use, attitude toward using, behavioural intention to use and actual system use of computers. The TAM predicted that perceived usefulness and ease of use are the two specific beliefs which are the most relevant to computer usage behaviours. On the other hand, the TRA is a well-known and often-used theory of human behaviour which consists of beliefs and evaluations, normative beliefs and motivations to comply, attitude toward behaviour, subjective norms, behavioural intention and actual behaviour, as indicated in Figure 1.2. However, these theories or models (TAM and TRA) are poorly theorised in educational psychology. This means that these models did not have psychological constructs such as computer self-efficacy and satisfaction.

Thus, this investigation has adopted an earlier model, the technology satisfaction model. This is because the TSM (Islam, 2014) has combined the TAM as well as educational psychology theories, which have been extensively applied by educational psychologists (Liu et al., 2018; Lodewyk, 2018; Phan

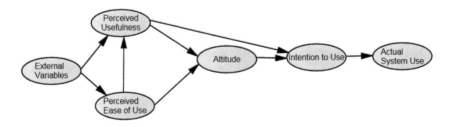

Figure 1.1 Technology acceptance model (adapted from Islam, 2014).

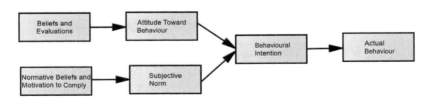

Figure 1.2 Theory of reasoned action (source: Davis et al., 1989).

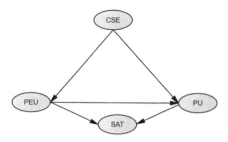

Figure 1.3 Technology satisfaction model (Islam, 2014).

Note: computer self-efficacy (CSE), satisfaction (SAT), perceived usefulness (PU) and perceived ease of use (PEU).

et al., 2018), such as social cognitive theory (Bandura, 1986). Self-efficacy is the key component of social cognitive theory, a theory which has been contributing widely in various arenas of education. The TSM has also included an intrinsic motivation component, namely "satisfaction". The TSM posits that users' satisfaction with technology is driven by three factors: the technology's perceived usefulness, its perceived ease of use, and the user's computer self-efficacy.

The TSM has proven "strong predictive power" in gauging students' satisfaction with new technology (Islam, 2014). Jiang et al. (2021, p. 16) claimed, "the findings obtained from the TSM have broadened the existing body of knowledge and current understanding of university students' satisfaction with using online learning platforms". Du et al. (2022, p. 7) stated, "through the application of TSM, which is considered to be an innovative model for examining learners' SAT, this study found that academic databases have been vital in helping postgraduate students accomplish their learning and research activities during the COVID-19 pandemic". Islam and Sheikh (2020, p. 1) asserted,

> the TSM shows that 69% of the variance in the postgraduate students' satisfaction can be explained by perceived usefulness, computer self-efficacy and perceived ease of use, which confirms that the TSM has a strong predictive power and is viable for measuring the endogenous variable. The findings also make a significant contribution to the testing of databases for academics, service providers and researchers in higher education.

However, it is not yet clear whether this model could be viable to assess factors underlying Chinese university lecturers' ICT satisfaction. On the other hand, the TSM literature indicates that it previously did not measure how lecturers' perceived ease of use (PEU) could have a direct impact on the perceived usefulness of ICT (PU). Subsequently, the TSM also did not explore how lecturers' computer self-efficacy (CSE) would have an indirect impact on perceived usefulness mediated by the perceived ease of use of ICT. As such, the present research includes these associations within the model as displayed in Figure 1.3. The associations between the features of the TSM are described in the next section to generate the present hypotheses, along with operational definitions of the terms that have been used with the model.

1.2.1 Hypotheses of the TSM

"Self-efficacy" refers to the extent to which a person has the belief that he or she can master a specific skill. Self-efficacy beliefs function as a vital set of proximal determinants of human action, motivation and affect, influencing action through cognitive, affective and motivational intervening processes (Bandura, 1989). In line with this definition, modern psychologists (Cheema & Skultety, 2017) have claimed that self-efficacy defines the belief of a person in his or her own capability and ability to succeed in accomplishing a given assignment. Thus, subject-specific self-efficacy means a measurement of self-confidence in a personal capacity to accomplish tasks associated with that subject. It is not necessary for this confidence to mirror actual ability in the subject: it can be an under- or over-estimation of true ability. As a result of emerging educational technology, researchers have introduced the idea of "computer self-efficacy", which refers to lecturers' confidence in their ability to use new technologies (Islam, 2015). Computer self-efficacy was found by many researchers to be associated with students' perceived ease of use and perceived usefulness of various ICT applications such as computer simulation (Liu & Huang, 2015), wireless internet (Islam, 2014; Islam et al., 2015; Islam et al., 2018), research databases (Islam & Sheikh, 2020; Chen et al., 2020; Du et al., 2022) and electronic textbooks (Chiu, 2017). However, this association has not been frequently demonstrated in Chinese university lecturers' ICT use. On the other hand, recent studies have claimed that self-esteem (Cheng, 2018) and social presence (Huang, 2017a) are significantly associated with perceived ease of use and usefulness. Due to these mixed associations, this investigation predicts that:

> H1: Chinese university lecturers' computer self-efficacy will be directly associated with the perceived ease of use of ICT applications.
> H2: Chinese university lecturers' computer self-efficacy will be directly associated with the perceived usefulness of ICT applications.

The TAM model (Davis et al., 1989) explored the idea that perceived ease of use and usefulness are the main components of user technology acceptance. Moreover, the TSM model (Islam, 2014) recently discovered that these two antecedents are solely powerful in measuring learner satisfaction. To explain these two factors, the present study reports their recent definitions as related to ICT. For instance, Islam (2015) suggested that "perceived usefulness" refers to the benefits perceived by lecturers through using ICT, while "perceived ease of use" means the ease of using ICT. Studies have identified these two influential dimensions as key in estimating learner satisfaction (Du et al., 2022; Jiang et al., 2022; Bin et al., 2020; Chen et al., 2020; Joo & Choi, 2016; Islam, 2014, 2015; Masrek & Gaskin, 2016; Islam & Sheikh, 2020; Islam, 2016; Islam et al., 2018; Islam et al., 2019; Li et al., 2021). However, the majority of these investigations have still claimed an association with either attitude (Cheng, 2018; Chiu, 2017; Liu & Huang, 2015; Huang, 2017b) or intention to use new technology (Poong et al., 2017). Against this background, the present research will validate similar associations which were generally in line with the findings of TSM in the case of Chinese lecturers' ICT satisfaction and predicts that:

> *H3*: Chinese university lecturers' satisfaction in using new technology will be directly associated with its perceived ease of use.
> *H4*: Chinese university lecturers' satisfaction in using new technology will be directly associated with its perceived usefulness.

Research related to technology acceptance has shown that learners' perceived ease of use is associated with the perceived usefulness of new technologies (Cheng, 2018; Poong et al., 2017; Chiu, 2017; Huang, 2017b; Yuan et al., 2016). Except for a few current studies (Huang, 2017b; Huang, 2015), such associations were generally in line with the results of the TAM (Davis et al., 1989), while the TSM did not include this relationship (see Figure 1.1). Therefore, the current research incorporates this association into the original TSM and predicts that:

> *H5*: Chinese university lecturers' perceived ease of use of the new technology will be directly associated with their perception of its usefulness.

Above all, the hypotheses have been constructed based on the direct associations of exogenous, mediating and endogenous variables of the TSM as discussed throughout extensive literature. Nevertheless, prior investigations have not adequately explored how mediating variables, such as perceived ease of use and usefulness, can establish indirect associations between either students' or lecturers' computer self-efficacy and their satisfaction with new technologies in higher education. For instance, Bin et al. (2020) found that technical and vocational college teachers' perception of self-efficacy and subsequent satisfaction with new technologies were indirectly influenced by the technologies' perceived usefulness and ease of use. Chen et al. (2020) claimed that the online database adoption and satisfaction (ODAS) model confirmed that the perceived ease of use and usefulness of databases are indirectly associated with postgraduate students' computer self-efficacy and satisfaction with the databases. According to Islam (2015), "lecturer satisfaction" is the degree to which the use of new technology is in line with lecturers' present values, needs and experiences. However, the TSM validated the mediating relationships among its constructs based on a student sample. Thus, the present research anticipates that these mediating variables will play the same role for Chinese lecturers and so assumes that:

> *H6*: Chinese university lecturers' perceived ease of use and usefulness will together mediate the indirect association between computer self-efficacy and satisfaction in using ICT.
>
> *H7*: Chinese university lecturers' perceived ease of use will mediate the indirect association between computer self-efficacy and perceived usefulness of ICT.
>
> *H8*: Chinese university lecturers' perceived usefulness will mediate the indirect association between perceived ease of use and satisfaction in using ICT.

1.3 Method

This study was undertaken at one public university in China. Before collecting the data, this researchers gained ethical clearance and informed consent from the participants. The investigation adopted a 56-item survey, originally developed by Islam (2015), which was administered to 200 lecturers. The authors selected lecturers from seven colleges of a university where 811 faculty members were employed. However, four respondents were not included in the final study due to their incomplete feedback. Before administering the survey, the questionnaire had gone through the translation-and-back-translation technique to assure that the meanings of the adapted English version of the instrument were not different in the translated Chinese questionnaire. Next, the questionnaire was systematically tested for face validity, and a pilot study using the Rasch model rating scale was carried out. In this empirical research, this researcher adapted the Rasch model for validating the psychometric properties of the items and persons involved. Through using the Rasch model, this study obtained a large number of valid items due to the minimum likelihood estimation. However, the majority of the studies related to educational technology have been ignored in order to validate the persons, who are important for the real measurement. A 56-item survey contained four domains, including the background information of Chinese lecturers, as shown in Table 1.1. The

Table 1.1 Components Measured in the Questionnaire

Facets	Likert Scale	No. of Items
Computer Self-Efficacy	1 to 7 (strongly disagree → strongly agree)	15
Perceived Ease of Use		15
Perceived Usefulness		16
Satisfaction	1 to 7 (very dissatisfied → very satisfied)	10
Total		*56*

8 *Applying the Rasch Model and Structural Equation Modeling to Higher Education*

Table 1.2 The Background Information of Chinese Lecturers

University	Variable	Category	% of Respondents
Public Universit	Gender	Male	48
		Female	52
		Total	100
	Age	25–29	8
		30–34	29
		35–39	21
		40–44	17
		Over 45	25
		Total	100
	Work experience	1–5	22
		6–10	32
		11–15	18
		16–20	10
		Over 21	18
		Total	100
	Education level	Undergraduate	19
		Master	48
		PhD	33
		Total	100
	Designation	Professor	7
		Associate Professor	38
		Senior Lecturer	2
		Lecturer	53
		Total	100

data from Chinese lecturers were collected using a stratified random sampling procedure and were analysed by applying structural equation modeling and the Rasch model.

One-hundred ninety-six useable Chinese lecturers' responses were collected from seven colleges of a public university in China, and they were divided into two groups according to gender. In this study, female lecturers (52%) were slightly more represented than male lecturers (48%). Most of the Chinese lecturers were aged between 30 and 44 years, and their years of experience ranged from 6 to 10 years, as indicated in Table 1.2. The majority of Chinese lecturers had master's degrees (48%), and they also held the position of lecturer (53%) during this study.

The Rasch model has been frequently applied in education, business, psychology, health and other social sciences. However, researchers have not yet developed and validated the instrument using the Rasch model, especially in ICT assessment in Chinese higher education. Hence, this research validated the technology satisfaction scale for investigating Chinese lecturers' satisfaction with new technologies. The scale's reliability (e.g., summary statistics and item polarity map) and validity (e.g., item fit order, item map and principal components) were tested using Rasch analysis through Winsteps version 3.49, and the detailed Rasch outputs are reported in Table 1.3. The results of this study

Table 1.3 Summary of Rasch Outputs

Summary Statistics		Item Polarity Map		Item Fit Order			Principal Components
Reliability Scores		PTMEA CORR.*	ITEMS	INFIT MNSQ	OUTFIT MNSQ	ITEMS	Variance
Items	.96	.43	grat5	1.39	1.62	peu11	65.8%
Persons	.97	.47	cse5	1.53	1.39	grat5	
Separation Scores		.47	peu1	1.21	1.48	cse14	
Items	4.73	.49	pu9	1.39	1.39	pu9	
Persons	5.39	.49	pu10	1.29	1.39	peu1	
		.50	cse15	1.20	1.38	pu10	
		.50	cse6	1.27	1.17	cse5	
		.50	peu11	1.07	1.26	peu2	
		.51	pu15	1.23	1.26	pu12	
		.51	peu2	1.24	1.24	pu14	
		.51	pu14	1.11	1.24	peu4	
		.52	cse14	1.22	1.08	cse6	
		.52	cse7	1.18	1.22	peu9	
		.53	cse11	1.17	1.21	peu10	
		.53	cse4	1.18	1.14	pu15	
		.54	cse1	1.15	1.09	grat10	
		.54	pu16	1.10	1.15	cse1	
		.54	peu4	1.12	1.09	grat7	
		.55	peu15	1.09	1.10	pu16	
		.55	grat7	1.05	1.09	peu8	
		.56	cse8	.96	1.07	pu11	
		.57	cse12	1.06	1.00	peu15	
		.57	cse3	1.01	1.06	peu13	
		.57	pu8	1.05	1.01	pu13	
		.57	cse10	1.04	.99	grat6	
		.58	cse9	1.03	.93	cse11	
		.58	peu9	.99	.96	grat3	
		.58	grat1	.98	.91	cse8	
		.58	cse13	.97	.90	grat9	
		.58	peu10	.96	.94	grat8	
		.59	peu3	.95	.96	peu3	
		.59	pu12	.94	.86	cse7	
		.59	pu13	.92	.85	grat2	
		.60	peu13	.88	.90	grat1	
		.61	grat3	.90	.90	cse4	
		.61	grat10	.89	.84	peu14	
		.61	pu11	.88	.86	grat4	
		.62	grat4	.84	.81	cse9	
		.62	peu8	.80	.84	peu5	
		.62	peu6	.83	.80	peu7	
		.62	grat6	.83	.76	cse12	
		.62	cse2	.83	.79	pu7	
		.63	grat9	.80	.82	pu1	
		.63	grat8	.77	.82	pu2	
		.63	grat2	.81	.73	cse10	

(*Continued*)

10 *Applying the Rasch Model and Structural Equation Modeling to Higher Education*

Table 1.3 (Continued)

Summary Statistics	Item Polarity Map		Item Fit Order			Principal Components
Reliability Scores	PTMEA CORR.*	ITEMS	INFIT MNSQ	OUTFIT MNSQ	ITEMS	Variance
.63	peu7		.80	.78	cse3	
.64	peu12		.77	.79	peu12	
.64	peu5		.78	.78	pu6	
.66	peu14		.76	.74	cse2	
.66	pu1		.70	.66	pu4	
.68	pu7		.66	.65	pu5	
.68	pu6		.61	.61	pu3	
.68	pu2					
.71	pu5					
.71	pu4					
.72	pu3					

*PTMEA CORR. = point measure correlation

supported the theoretical structure of the psychometric properties as involving computer self-efficacy, satisfaction, perceived ease of use and usefulness of ICT. Fifty-six items fit the model, and the scales had good psychometric properties. The proportion of variance explained by the measures was 65.8%, which indicated that the items were able to reflect the Chinese lecturers' satisfaction with new technologies. As exhibited in the item map (see Figure 1.4), Chinese lecturers are allocated on the left side of the scale, and items are allocated on the right side. Lecturers with perceived higher ability in using ICT in terms of their computer self-efficacy, satisfaction, perceived usefulness and ease of use are located higher on the scale. On the other hand, the items which were more complicated for the lecturers to agree on are located higher on the scale. The item map indicates that, overall, Chinese lecturers' abilities were higher on the scale than the items' difficulty, and thus, in general, lecturers tended to agree that providing ICT facilities was satisfactory for teaching and research purposes in their university. Comparably few lecturers (located below items peu4 and peu6) did not agree that their university provide adequate ICT facilities to support their teaching and research activities. Therefore, some difficult items were required in order to measure those lecturers with strong satisfaction. For this reason, inclusion of more difficult items was recommended for future studies.

Meanwhile, the outfit mean square score for item peu11 and the infit mean square score for item grat5 were slightly larger than the suggested range of 0.5 to 1.5 (Bond & Fox, 2001) as displayed in Table 1.3. As such, these two items were excluded when conducting further analyses (see Appendix).

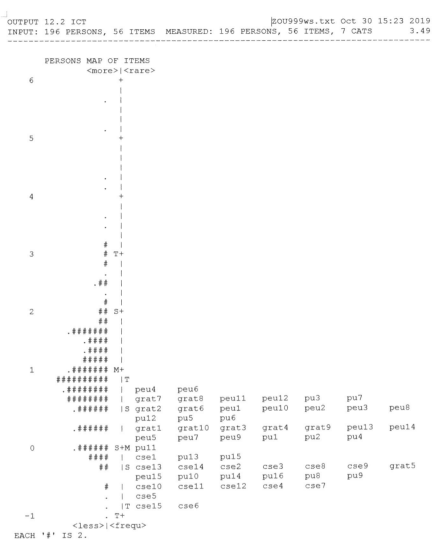

Figure 1.4 Item map.

Table 1.4 summarises 54 valid indicators obtained through Rasch analysis. These indicators are measuring four constructs such as computer self-efficacy (CSE), satisfaction (SAT), perceived usefulness (PU) and perceived ease of use (PEU).

Table 1.4 The Summary of the Indicators

Constructs	Valid Items		α
PEU	PEU1	I find the university ICT facilities easy to use.	.929
	PEU2	I find it easy to access the university research databases.	
	PEU3	It is easy for me to become skillful at using university databases for conducting research.	
	PEU4	Interacting with the research databases system requires minimal effort on my part.	
	PEU5	I find it easy to get the research databases to help facilitate my research.	
	PEU6	Interacting with the university research databases system is very stimulating for me.	
	PEU7	I find it easy to select articles/journals of different categories using research databases (e.g., education, engineering and business).	
	PEU8	With Wireless Internet, I find it easy to access university databases to do research.	
	PEU9	Wireless Internet allows me to access research and learning materials from Web Browser (WWW).	
	PEU10	Wireless Internet is easy to use for teaching and research.	
	PEU12	My interaction with university ICT services available is clear and understandable.	
	PEU13	I find it easy to download teaching, learning and research materials using Wireless Internet in terms of speed.	
	PEU14	It is easy for me to use multimedia facilities at the university.	
	PEU15	I find it easy to use Microsoft Office for teaching and research purposes.	
PU	PU1	Using the ICT facilities at the university enables me to accomplish tasks more quickly.	.943
	PU2	Using the ICT services available at the university increases my research productivity.	
	PU3	The current university ICT system makes work more interesting.	
	PU4	ICT facilities at the university improve the quality of the work I do.	
	PU5	Using the university ICT facilities enhances my research skills.	
	PU6	Using the university ICT facilities would make it easier for me to find information.	
	PU7	ICT services at the university make information always available to users.	
	PU8	Using the ICT facilities helps me write my journal articles.	
	PU9	My job would be difficult to perform without the ICT facilities.	
	PU10	Using the ICT facilities saves my time.	
	PU11	Using the university ICT facilities provides me with the latest information on specific areas of research.	
	PU12	I can use ICT facilities from anywhere, anytime at the campus.	
	PU13	I believe that the use of ICT in the classroom enhances student learning in my discipline.	

	PU14	I believe that e-mail and other forms of electronic communication are important tools in faculty/student communication.	
	PU15	I believe that web-based instructional materials enhance student learning.	
	PU16	Using the ICT services enables me to download teaching, learning and research materials from the internet.	
CSE	CSE1	I have the skills required to use computer applications for writing my research papers.	.958
	CSE2	I have the skills and knowledge required to use computer applications for demonstrating specific concepts in class.	
	CSE3	I have the skills required to use computer applications for presenting lectures.	
	CSE4	I have the skills required to communicate electronically with my colleagues and students.	
	CSE5	I have the ability to e-mail, chat, download teaching, learning and research materials, and search different websites.	
	CSE6	I have the ability to use the Wireless Internet service provided by the university.	
	CSE7	I have the skills required to use ICT facilities to enhance the effectiveness of my teaching, learning and research.	
	CSE8	I feel capable of using the university research databases for writing journal papers.	
	CSE9	I have the ability to navigate my way through the ICT facilities.	
	CSE10	I have the skills required to use the ICT facilities to enhance the quality of my research works.	
	CSE11	I have the ability to save and print journals/articles from the research databases.	
	CSE12	I have the knowledge and skills required to benefit from using the university ICT facilities.	
	CSE13	I am capable of accessing the research databases from the university website.	
	CSE14	I am capable of downloading and installing research-related software using ICT facilities at the university.	
	CSE15	I am capable of using multimedia facilities for delivering lecturers.	
SAT	GRAT1	Overall, I am satisfied with the ease of completing my task using ICT facilities.	.947
	GRAT2	I am satisfied with the ICT facilities provided by the university.	
	GRAT3	I am satisfied with the ease of use of the ICT facilities.	
	GRAT4	The university ICT facilities have greatly affected the way I search for information and conduct my research.	
	GRAT6	Overall, I am satisfied with the amount of time it takes to complete my task.	
	GRAT7	I am satisfied with the structure of accessible information (available as categories of research domain, or by date of issue – of journals in particular, or as full-texts or abstracts of theses and dissertations) of the university research databases.	
	GRAT8	I am satisfied with the support information provided by the university's ICT facilities.	
	GRAT9	I am satisfied in using ICT facilities for teaching, learning and research.	
	GRAT10	Overall, I am satisfied with the Wireless Internet service provided at the university.	

14 *Applying the Rasch Model and Structural Equation Modeling to Higher Education*

1.4 Findings

A 54-item instrument was validated through the Rasch model and was considered valid to assess four facets of the measurement model, namely, computer self-efficacy, satisfaction, perceived usefulness and perceived ease of use using confirmatory factor analysis (CFA) in regards to their convergent and discriminant validity. Firstly, a four-facet measurement model was evaluated using a few fit statistics such as chi-square/degree of freedom ($\chi^2/df \leq 3$), root mean square error of approximation (RMSEA $\leq$ .08 or .1), comparative fit index (CFI $\geq$ .90), and Tucker–Lewis index (TLI $\geq$ .90) based on the recommendation of statisticians who suggested these indices (Hu & Bentler, 1999). However, according to the results of CFA, the measurement model did not fit the empirical data as indicated in Figure 1.5: $\chi^2 = 4349.551$; $df = 1371$; $p = 0.000$; RMSEA $= 0.106$; CFI $= 0.706$; and TLI $= 0.693$. The output of AMOS version 18 for the measurement model showed that several indicators were loaded on each other as contradicting the measurement criteria. Against this backdrop, the model needs to be re-specified, while all the indicators were significant.

The four-facet re-specified model fit the empirical data well: $\chi^2 = 249.122$; $df = 129$; $p = 0.000$; RMSEA $= 0.069$; CFI $= 0.953$; and TLI $= 0.944$. This is because several items were excluded from the constructs of the re-specified model one at a time (see Table 1.5) based on their modification indices, such as computer self-efficacy (CSE1, CSE3, CSE4, CSE5, CSE6, CSE7, CSE8, CSE11, CSE13 and CSE14), perceived ease of use (PEU1, PEU2, PEU7, PEU8, PEU9, PEU10, PEU12, PEU13, PEU14 and PEU15), satisfaction (SAT1, SAT3 and SAT8), and perceived usefulness (PU1, PU3, PU4, PU7, PU8, PU9, PU10, PU11, PU12, PU13, PU14, PU15 and PU16). These items indicate multicollinearity.

All these parameters had to be excluded due to the model's fitness and multicollinearity, as well as convergent and discriminant validity. Furthermore, the correlation matrix confirmed that there is no indication of multicollinearity among the indicators as none of coefficients exceeded the cut-off point of 0.85 (Fornell & Larcker, 1981). The four-facet re-specified model revealed a significant level of loadings ranging from 0.72 to 0.89 for all the constructs and their interrelationships, as shown in Figure 1.6.

This research obtained the values of correlations between the dimensions and standardised regression weights through AMOS output to calculate the convergent and discriminant validity. Table 1.6 exhibits the composite reliability (CR) and average variance extracted (AVE) scores related to convergent validity (Hair et al., 2010; Fornell & Larcker, 1981) where all the CR and AVE scores are above 0.853 and 0.592, respectively.

Similarly, the square root of the AVE for PEU, CSE, SAT and PU is larger than the interrelationships as bolded in Table 1.6. For instance, PEU <--> CSE ($\beta = 0.402$; $p < .000$; CR $= 4.419$); PEU <--> SAT ($\beta = 0.616$;

Assessment of ICT in Higher Education Applying the TSM   15

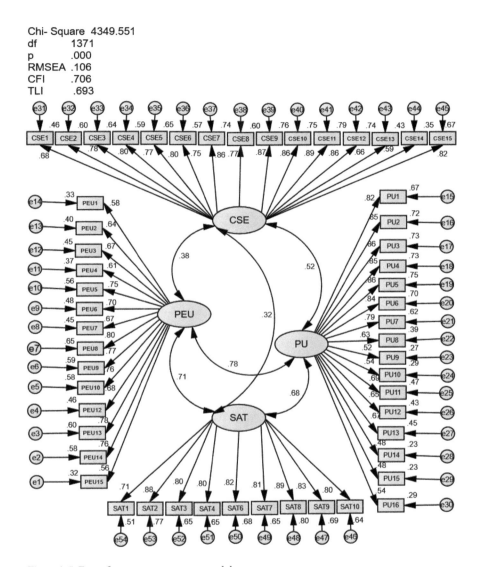

Figure 1.5 Four-factor measurement model.

Note: computer self-efficacy (CSE), satisfaction (SAT), perceived usefulness (PU) and perceived ease of use (PEU).

$p < .000$; CR = 6.190); PEU <--> PU (β = 0.760; $p < .000$; CR = 4.673); CSE <--> SAT (β = 0.298; $p < .000$; CR = 3.609); CSE <--> PU (β = 0.423; $p < .000$; CR = 4.673); SAT <--> PU (β = 0.679; $p < .000$; CR = 6.845).

16   *Applying the Rasch Model and Structural Equation Modeling to Higher Education*

Table 1.5 The List of Deleted Items

Item Deleted	Chi-Square (χ^2)	df	p Value	RMSEA	CFI	TLI
CSE13	4102.465	1319	.000	.104	.717	.705
PU9	3837.069	1268	.000	.102	.732	.719
PU14	3631.221	1218	.000	.101	.743	.731
CSE1	3443.487	1169	.000	.100	.751	.739
SAT8	3288.972	1121	.000	.100	.752	.740
PU1	3154.172	1074	.000	.100	.754	.742
PU8	2960.904	1028	.000	.098	.765	.753
PU10	2769.054	983	.000	.097	.778	.766
PU15	2624.939	939	.000	.096	.787	.775
PEU9	2499.643	896	.000	.096	.791	.779
PEU10	2384.894	854	.000	.096	.795	.783
PU13	2247.740	813	.000	.095	.803	.792
PU16	2099.139	773	.000	.094	.814	.803
PU11	1976.176	734	.000	.093	.822	.811
SAT3	1865.955	696	.000	.093	.826	.814
CSE5	1717.247	659	.000	.091	.835	.824
PEU13	1633.877	623	.000	.091	.838	.826
PEU2	1520.900	588	.000	.090	.846	.835
PEU7	1390.040	554	.000	.088	.857	.846
CSE3	1284.315	521	.000	.087	.863	.853
CSE4	1202.847	489	.000	.087	.868	.857
PU3	1127.402	458	.000	.087	.868	.857
CSE6	1041.803	428	.000	.086	.874	.863
PU7	950.639	399	.000	.084	.880	.869
PEU15	856.498	371	.000	.082	.891	.881
PEU14	777.705	344	.000	.080	.899	.889
PEU8	720.541	318	.000	.081	.903	.893
SAT1	636.575	294	.000	.077	.914	.904
CSE8	578.653	270	.000	.077	.918	.909
CSE11	528.755	247	.000	.076	.919	.909
CSE7	490.651	225	.000	.078	.918	.908
PU12	421.614	205	.000	.074	.930	.922
PEU12	378.870	185	.000	.073	.935	.927
CSE14	313.512	166	.000	.068	.949	.941
PU4	286.702	148	.000	.069	.947	.939
PEU1	249.144	130	.000	.069	.953	.945

These results suggest conducting further estimation on the technology satisfaction model, which is the structural model of this research. Table 1.7 contains detailed information on the valid items of the four-facet re-specified measurement model.

Structural equation modeling is one of the strongest and most frequently applied analytical tools for validating and estimating the measurement and structural models using the maximum likelihood estimation procedure (Byrne, 2000). The remaining 18 items for four constructs, namely CSE, SAT, PU and PEU, validated in the measurement model have been

Assessment of ICT in Higher Education Applying the TSM 17

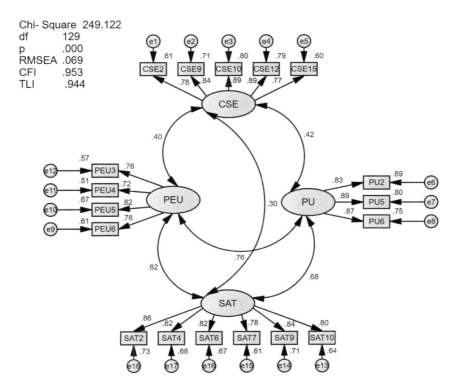

Figure 1.6 The re-specified four-factor measurement model.

Table 1.6 The Scores of the Convergent and Discriminant Validity

Facets	CR	AVE	PEU	CSE	SAT	PU
PEU	0.853	0.592	**0.769**			
CSE	0.922	0.702	0.402	**0.838**		
SAT	0.925	0.674	0.616	0.298	**0.821**	
PU	0.899	0.747	0.760	0.423	0.679	**0.864**

redesigned to evaluate the TSM and its proposed six hypotheses. The findings of the TSM attested that all the hypotheses were accepted as indicated in Figure 1.7. For example, Chinese university lecturers' CSE had a direct association with their PEU (β = 0.40; p = .000; CR = 5.009) and PU (β = 0.14; p = .034; CR = 2.123) of ICT (*H1* and *H2*). Chinese lecturers' PEU (β = 0.24; p = .033; CR = 2.131) and PU (β = 0.50; p = .000; CR = 4.485) also had a direct association with their SAT in using ICT (*H3* and *H4*). Likewise, Chinese university lecturers' PEU was directly associated

Table 1.7 The Detailed Information on the Valid Items of the Revised Measurement Model

Parameters of the Facets	Loadings	M	SD	α
PEU				
PEU3 It is easy for me to become skilful at using university databases for conducting research.	0.76	5.01	1.236	
PEU4 Interacting with the research databases system requires minimal effort on my part.	0.72	4.71	1.320	0.852
PEU5 I find it easy to get the research databases to help facilitate my research.	0.82	5.11	1.229	
PEU6 Interacting with the university research databases system is very stimulating for me.	0.78	4.65	1.333	
CSE				
CSE2 I have the skills and knowledge required to use computer applications for demonstrating specific concepts in class.	0.78	5.56	1.044	
CSE9 I have the ability to navigate my way through the ICT facilities.	0.84	5.60	1.046	0.919
CSE10 I have the skills required to use the ICT facilities to enhance the quality of my research works.	0.89	5.68	.989	
CSE12 I have the knowledge and skills required to benefit from using the university ICT facilities.	0.89	5.78	.965	
CSE15 I am capable of using multimedia facilities for delivering lecturers.	0.77	5.97	.955	
SAT				
SAT2 I am satisfied with the ICT facilities provided by the university.	0.86	4.93	1.361	
SAT4 The university ICT facilities have greatly affected the way I search for information and conduct my research.	0.82	5.18	1.276	
SAT6 Overall, I am satisfied with the amount of time it takes to complete my task.	0.82	4.94	1.413	0.924
SAT7 I am satisfied with the structure of accessible information (available as categories of research domain, or by date of issue – of journals in particular, or as full-texts or abstracts of theses and dissertations) of the university research databases.	0.78	4.87	1.358	
SAT9 I am satisfied in using ICT facilities for teaching, learning and research.	0.84	5.21	1.315	
SAT10 Overall, I am satisfied with the Wireless Internet service provided at the university.	0.80	5.14	1.443	
PU				
PU2 Using the ICT services available at the university increases my research productivity.	0.83	5.14	1.317	
PU5 Using the university ICT facilities enhances my research skills.	0.89	5.06	1.237	0.898
PU6 Using the university ICT facilities would make it easier for me to find information.	0.87	5.05	1.323	

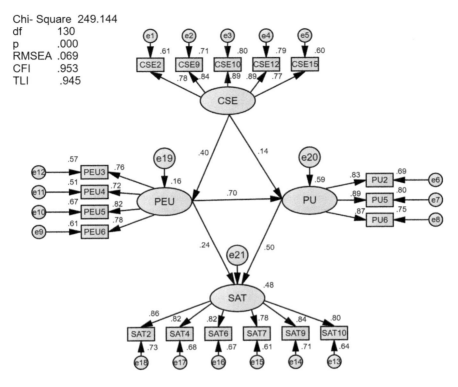

Figure 1.7 The technology satisfaction model.

with their PU (β = 0.70; *p* = .000; CR = 8.436) of ICT (*H5*). On the other hand, Chinese academicians' PEU (chi-square, χ^2 = 1.735; *p* < 0.05) had only mediated the indirect association between CSE and SAT instead of PU. Thus, this hypothesis was partially validated (*H6*). Chinese lecturers' PEU (chi-square, χ^2 = 3.517; *p* < 0.001) had also mediated the indirect association between CSE and PU of ICT (*H7*). Finally, Chinese academicians' PU (chi-square, χ^2 = 3.739; *p* < 0.001) had significantly mediated the indirect association between PEU and SAT in using of ICT (*H8*). These indirect associations were investigated using the Sobel test (Sobel, 1982).

Additionally, as shown in Table 1.8, bootstrap provided evidence of a 95% confidence interval for the significant hypothesised direct and indirect paths. The significant direct path from PEU to PU had a large effect size, while the direct paths from CSE to PEU and PU to SAT had medium effect sizes. The direct paths from CSE to PU and PEU to SAT had small effect sizes. On the other hand, the significant indirect paths from CSE to SAT (mediated by PEU) and PEU to SAT (mediated by PU) had large effect sizes. Moreover, the indirect path from CSE to PU (mediated by PEU) had a small effect size.

20 *Applying the Rasch Model and Structural Equation Modeling to Higher Education*

Table 1.8 Confidence Interval of Direct and Indirect Effects

Confidence Intervals		Lower Bounds	Upper Bounds	Two-Tailed Significance
Direct Effects	CSE→PEU	.324	.716	.001
	CSE→PU	.008	.386	.038
	PEU→PU	.571	.925	.001
	PEU→SAT	.064	.654	.033
	PU→SAT	.175	.861	.006
Indirect Effects	CSE→PU	.255	.546	.001
	CSE→SAT	.305	.620	.000
	PEU→SAT	.190	.625	.004

Table 1.9 The Detailed Information of Effect Sizes, Summary of Hypotheses and Variances

Hypotheses			Path Coefficients (β)	Results
H1: CSE→PEU			.40	Accepted
H2: CSE→PU			.14	Accepted
H3: PEU→SAT			.24	Accepted
H4: PU→SAT			.50	Accepted
H5: PEU→PU			.70	Accepted
H6: CSE→SAT (PEU)			.132 (≥0.080)	Accepted
H6: CSE→SAT (PU)			.100 (≥0.080)	Rejected
H7: CSE→PU (PEU)			.377 (≥0.080)	Accepted
H8: PEU→SAT (PU)			.392 (≥0.080)	Accepted
Standardised Effect Sizes			Variances	Percentages
Direct	CSE→PEU	0.402	PU (CSE)	59%
	CSE→PU	0.139	PEU (CSE)	16%
	PEU→PU	0.704	SAT (CSE, PU and PEU)	48%
	PEU→SAT	0.236		
	PU→SAT	0.499		
Indirect	CSE→PEU→PU	0.283		
	CSE→PEU→SAT	0.305		
	PEU→PU→SAT	0.351		

The standardised direct and indirect effect sizes of the TSM along with its variance and summary of hypotheses are reported in Table 1.9.

1.5 Discussion

The outcomes of assessment in regards to ICT presented in this study have theoretically, practically and methodologically contributed to the body of scientific knowledge related to various aspects of education. The results of the present research prove that Chinese university lecturers' computer ability is the main factor in perceiving ICT as useful, and lecturers do not face difficulties in using it for their academic purposes. Hence, these findings suggest that academics and researchers should consider advancing lecturers' computer self-efficacy. Thus, they can achieve more benefits through ICT applications and confront less complexity in accomplishing their work, as it could be performed through the effective training courses or workshops. In line with the results of this research, investigations found that learners' computer self-efficacy was associated with the perceived ease of use and usefulness of various ICT applications (Chiu, 2017; Liu & Huang, 2015; Islam, 2014). However, Cheng (2018) and Huang (2017a) asserted that learners' self-esteem and social presence were associated with perceived usefulness and ease of use.

This research confirmed that the mediating variables of TSM (Islam, 2014), such as perceived ease of use and usefulness, are the important direct determinants of Chinese lecturers' satisfaction, while the TAM (Davis et al., 1989), along with other researchers, stated that they are valid antecedents of technology acceptance (Joo & Choi, 2016; Masrek & Gaskin, 2016). These results provide a strong recommendation for academicians and service providers. As an example, lecturers' satisfaction greatly depend on their experiences while using new technology. As such, to extend lecturers' satisfaction, the authority of service providers should certify that high-quality ICT facilities are available within the academic environment so that lecturers can gain desirable advantages and will not perceive difficulties in performing their academic-related jobs. Along these lines, academicians should also conduct ongoing assessment in regards to the quality of ICT services so that better teaching, learning and research environments will be implemented in tertiary education. However, researchers have still identified that the TAM variables are associated with not only attitude (Cheng, 2018; Chiu, 2017; Liu & Huang, 2015; Huang, 2017b) but also intention to use technology (Poong et al., 2017). Meanwhile, the relationship between perceived ease of use and perceived usefulness has been widely expounded upon in the literature.

This study also explored how Chinese university lecturers' perception of the usefulness of ICT is affected by their experiences in terms of accessibility, such as how easy it is to use: an outcome that is consistent with the TAM-based (Davis et al., 1989) result including new studies (Cheng, 2018; Poong et al., 2017; Chiu, 2017; Huang, 2017b; Yuan et al., 2016). On the other hand, in the previous study, the TSM (Islam, 2014) did not recognise the significance of this result, which can provide insights for future researchers to identify the strategies of getting more benefits through ICT.

Finally, the results related to mediating effects of perceived ease of use and usefulness generated a new platform for future researchers. For instance, the TSM (Islam, 2014) previously claimed that these mediating variables are helping learners to increase satisfaction because of their computer self-efficacy, which is applicable for this study in measuring Chinese lecturers' satisfaction. These findings showed multiply mediating associations between lecturers' computer ability and their satisfaction in regards to ICT use and acknowledged that lecturers' satisfaction is also dependent on their beliefs in ICT ability. So it is imperative for future researchers to take necessary initiatives to determine the various ways of improving their ICT skills. Due to the new relationship between perceived ease of use and usefulness, this research also indicates the indirect associations between lecturers' self-efficacy and usefulness along with their ease of use and satisfaction.

1.6 Conclusion

Assessment is an ongoing process in identifying problems and proposing possible solutions for studies concerned with extending the quality of teaching, learning and research on applying appropriate psychological or analytical tools. The combination of SEM and the Rasch model has been acknowledged as the real measurement for this study. The joint findings of these models suggest that the Rasch model explores a total of 54 valid items for validating all the constructs of the TSM due to the likelihood estimation and recommends them to be adopted for future research, although SEM analysis excluded several indicators for validating the measurement and structural models due to the maximum likelihood evaluation. Subsequently, advanced statistical analyses of the results revealed that the technology satisfaction model has "strong predictive power" in gauging Chinese lecturers' satisfaction with ICT. For instance, Chinese lecturers' satisfaction with ICT use was explored through the three underlying antecedents of their computer self-efficacy, the technology's perceived usefulness and ease of use, while computer self-efficacy was evident as being a powerful exogenous variable in assessing direct and indirect causal associations of the dimensions of TSM.

1.6.1 Implications for Policy and Practice

Assessment of Chinese lecturers' ICT use for their teaching and research purposes as well as applying structural equation modeling and the Rasch model will be essential for future researchers, academicians and practitioners in performing technological measurements in all arenas of tertiary education. Nonetheless, the TSM found that Chinese university lecturers' computer self-efficacy could only explain 16% of the variability of the ease of use of ICT. This result strongly recommends that policymakers prioritise the extension of lecturers' required ICT skills before providing extensive new technologies. Interestingly, the relationship between perceived ease of use and perceived usefulness was found to be strongest in TSM. This was addressed extensively in the literature. Based on the current findings, it could be concluded that the research appears to be well constructed and sound and that the results will likely be beneficial for future researchers who wish to assess Chinese lecturers' or Chinese college and university students' satisfaction with technology. These findings may encourage policymakers to establish a short course or ongoing workshops for faculty members who will be involved in hands-on practice. These courses or workshops would focus on analytical tools like the Rasch model and structural equation modeling with the goal of helping them to improve the productivity of their research and publication. To provide a better understanding of TSM and to examine its viability, regional or cross-cultural comparative investigations are recommended. This study assumes that there may be an interaction of the results and the participants' age and gender. Due to the small sample size, researchers did not explore the moderating effects of these variables, but they suggest that these effects be explored in future research. This study contributes to relevance in advancing knowledge in the field of ICT assessment. "Satisfaction" is a complicated concept to define, and this helps to provide a model for doing so.

1.7 Acknowledgements

This work was supported by the Peak Discipline Construction Project of Education at East China Normal University and Fundamental Research Funds for the Central Universities (2020ECNU-HLYT035).

1.8 References

Al-Azawei, A., Parslow, P., & Lundqvist, K. (2017). Investigating the effect of learning styles in a blended e-learning system: An extension of the technology acceptance model (TAM). *Australasian Journal of Educational Technology*, *33*(2), 1–23.

Areepattamannil, S., & Khine, M. S. (2017). Early adolescents' use of information and communication technologies (ICTs) for social communication in 20 countries: Examining the roles of ICT-related behavioral and motivational characteristics. *Computers in Human Behavior*, *73*, 263–272. http://dx.doi.org/10.1016/j.chb.2017.03.058

Baars, M., Leopold, C., & Paas, F. (2018). Self-explaining steps in problem-solving tasks to improve self-regulation in secondary education. *Journal of Educational Psychology*, *110*(4), 578–595. http://dx.doi.org/10.1037/edu0000223

Bandura, A. (1986). *Social foundations of thought and action: A social cognitive theory*. Prentice-Hall.

Bandura, A. (1989). Human agency in social cognitive theory. *American Psychologist*, *44*(9), 1175–1184. https://doi.org/10.1037/0003-066X.44.9.1175

Bennett, S., Dawson, P., Bearman, M., Molloy, E., & Boud, D. (2017). How technology shapes assessment design: Findings from a study of university teachers. *British Journal of Educational Technology*, *48*(2), 672–682. https://doi.org/10.1111/bjet.12439

Bin, E., Islam, A. Y. M. A., Gu, X., Spector, J. M., & Wang, F. (2020). A study of Chinese technical and vocational college teachers' adoption and gratification in new technologies. *British Journal of Educational Technology*, *51*(6), 2359–2375. https://doi.org/10.1111/bjet.12915

Bond, T. G., & Fox, C. M. (2001). *Applying the Rasch model: Fundamental measurement in the human science*. Lawrence Erlbaum.

Buuren, N. V., & Eggen, T. H. J. M. (2017). Latent-class-based item selection for computerized adaptive progress tests. *Journal of Computerized Adaptive Testing*, *5*(2), 22–43. https://doi.org/10.7333/1704–0502022

Byrne, B. M. (2000). *Structural equation modeling with Amos: Basic concepts, applications and programming*. Lawrence Erlbaum.

Campbell, C., & Bond, T. (2017). Investigating young children's human figure drawings using Rasch analysis. *Educational Psychology*, *37*(7), 888–906. https://doi.org/10.1080/01443410.2017.1287882

Cheema, J. R., & Skultety, L. S. (2017). Self-efficacy and literacy: A paired difference approach to estimation of over-/under-confidence in mathematics- and science-related tasks. *Educational Psychology*, *37*(6), 652–665. https://doi.org/10.1080/01443410.2015.1127329

Chen, H., Islam, A. Y. M. A., Gu, X., Teo, T., & Peng, Z. (2020). Technology-enhanced learning and research using databases in higher education: The application of the ODAS model. *Educational Psychology*, *40*(9), 1056–1075. https://doi.org/10.1080/01443410.2019.1614149

Cheng, E. W. L. (2018). Choosing between the theory of planned behavior (TPB) and the technology acceptance model (TAM). *Educational Technology Research and Development*. https://doi.org/10.1007/s11423-018-9598-6

Chiu, T. K. F. (2017). Introducing electronic textbooks as daily-use technology in schools: A top-down adoption process. *British Journal of Educational Technology*, *48*(2), 524–537. https://doi.org/10.1111/bjet.12432

Coniam, D., & Yan, Z. (2016). A comparative picture of the ease of use and acceptance of onscreen marking by markers across subject areas. *British Journal of Educational Technology*, *47*(6), 1151–1167. https://doi.org/10.1111/bjet.12294

Davis, F. D., Bagozzi, R. P., & Warshaw, P. R. (1989). User acceptance of computer-technology: A comparison of two theoretical models. *Management Science*, *35*(8), 982–1003.

Du, H., Chen, H., & Islam, A. Y. M. A. (2022). Students' perception of academic databases as recognition of learning and research during the COVID-19 pandemic. *Journal of Information Science*. https://journals.sagepub.com/doi/full/10.1177/01655515221118666

Farrell, T., & Rushby, N. (2016). Assessment and learning technologies: An overview. *British Journal of Educational Technology*, *47*(1), 106–120. https://doi.org/10.1111/bjet.12348

Fathali, S., & Okada, T. (2018). Technology acceptance model in technology-enhanced OCLL contexts: A self-determination theory approach. *Australasian Journal of Educational Technology*, *34*(4), 138–154. https://doi.org/10.14742/ajet.3629

Fishbein, M., & Ajzen, I. (1975). *Belief, attitude, intention and behavior: An introduction to theory and research*. Addison-Wesley.

Fornell, C., & Larcker, D. F. (1981). Evaluating structural equation models with unobservable variables and measurement error. *Journal of Marketing Research*, *48*, 39–50.

Gibbons, R. E., Xu, X., Villafane, S. M., & Raker, J. R. (2018). Testing a reciprocal causation model between anxiety, enjoyment and academic performance in postsecondary organic chemistry. *Educational Psychology*. https://doi.org/10.1080/01443410.2018.1447649

Graham, S. (2018). Race/ethnicity and social adjustment of adolescents: How (Not if) school diversity matters. *Educational Psychologist*, *53*(2), 64–77. https://doi.org/10.1080/00461520.2018.1428805

Hair, J. F. J., Black, W. C., Babin, B. J., & Anderson, R. E. (2010). *Multivariate data analysis a global perspective*. Pearson Education, Inc.

Halberstadt, A. G., Castrob, V. L., Chuc, Q., Lozadad, F. T., & Sims, C. M. (2018). Preservice teachers' racialized emotion recognition, anger bias, and hostility attributions. *Contemporary Educational Psychology*, *54*, 125–138. https://doi.org/10.1016/j.cedpsych.2018.06.004

Holzinger, A., Searle, G., & Wernbacher, M. (2011). The effect of previous exposure to technology (PET) on acceptance and its importance in usability engineering. *Universal Access in the Information Society International Journal*, *10*(3), 245–260. https://doi.org/10.1007/s10209-010-0212-x

Hu, L. T., & Bentler, P. M. (1999). Cutoff criteria for fit indexes in covariance structure analysis: Conventional criteria versus new alternatives. *Structural Equation Modeling*, *6*, 1–55.

Huang, T. K. (2015). Exploring the antecedents of screenshot-based interactions in the context of advanced computer software learning. *Computers & Education, 80*, 95–107. http://dx.doi.org/10.1016/j.compedu.2014.08.011

Huang, Y. (2017a). Exploring students' acceptance of team messaging services: The roles of social presence and motivation. *British Journal of Educational Technology, 48*(4). 1047–1061. https://doi.org/10.1111/bjet.12468

Huang, Y. M. (2017b). Exploring the intention to use cloud services in collaboration contexts among Taiwan's private vocational students. *Information Development, 33*(1), 29–42. https://doi.org/10.1177/0266666916635223

Huebner, A., Finkelman, M. D., & Weissman, A. (2018). Factors affecting the classification accuracy and average length of a variable-length cognitive diagnostic computerized test. *Journal of Computerized Adaptive Testing, 6*(1), 1–14. https://doi.org/10.7333/1802–060101

Islam, A. Y. M. A. (2014). Validation of the technology satisfaction model (TSM) developed in higher education: The application of structural equation modeling. *International Journal of Technology and Human Interaction, 10*(3), 44–57. https://doi.org/10.4018/ijthi.2014070104

Islam, A. Y. M. A. (2015). *Development and validation of the technology adoption and gratification (TAG) model in assessing teachers' ICT use* [Unpublished PhD dissertation, Institute of Graduate Studies, University of Malaya].

Islam, A. Y. M. A. (2016). Development and validation of the technology adoption and gratification (TAG) model in higher education: A cross-cultural study between Malaysia and China. *International Journal of Technology and Human Interaction, 12*(3), 78–105. https://doi.org/10.4018/IJTHI.2016070106

Islam, A. Y. M. A., Leng, C. H., & Singh, D. (2015). Efficacy of the technology satisfaction model (TSM): An empirical study. *International Journal of Technology and Human Interaction, 11*(2), 45–60. https://doi.org/10.4018/ijthi.2015040103

Islam, A. Y. M. A., Mok, M. M. C., Gu, X., Spector, J. M., & Leng, C. H. (2019). ICT in higher education: An exploration of practices in Malaysian universities. *IEEE Access, 7*(1), 16892–16908. https://doi.org/10.1109/ACCESS.2019.2895879

Islam, A. Y. M. A., Mok, M. M. C., Xiuxiu, Q., & Leng, C. H. (2018). Factors influencing students' satisfaction in using wireless internet in higher education: Cross-validation of TSM. *The Electronic Library, 36*(1), 2–20. https://doi.org/10.1108/EL-07-2016-0150

Islam, A. Y. M. A., & Sheikh, A. (2020). A study of the determinants of postgraduate students' satisfaction in using online research databases. *Journal of Information Science, 46*(2), 273–287. https://doi.org/10.1177/0165551519834714

Jiang, H., Islam, A. Y. M. A., Gu, X., & Spector, J. M. (2021). Online learning satisfaction in higher education during the COVID-19 pandemic: A regional comparison between eastern and western Chinese universities. *Education and Information Technologies, 26*(6), 6747–6769. https://doi.org/10.1007/s10639-021-10519-x

Jiang, H., Islam, A. Y. M. A., Gu, X., & Spector, J. M. (2022). Technology-enabled e-learning platforms in Chinese higher education during the pandemic age of COVID-19. *SAGE Open, 12*(2), 1–15. https://doi.org/10.1177/21582440221095085

Joo, S., & Choi, N. (2016). Understanding users' continuance intention to use online library resources based on an extended expectation-confirmation model. *The Electronic Library, 34*(4), 554–571. http://dx.doi.org/10.1108/EL-02-2015-0033

Lee, M., & Vlack, S. V. (2018). Teachers' emotional labour, discrete emotions, and classroom management self-efficacy. *Educational Psychology, 38*(5), 669–686. https://doi.org/10.1080/01443410.2017.1399199

Li, A., Islam, A. Y. M. A., & Gu, X. (2021). Factors engaging college students in online learning: An investigation of learning stickiness. *SAGE Open*, *11*(4), 1–15. https://doi.org/10.1177/21582440211059181

Liu, C. H., & Huang, Y. M. (2015). An empirical investigation of computer simulation technology acceptance to explore the factors that affect user intention. *Universal Access in the Information Society*, *14*(3), 449–457. https://doi.org/10.1007/s10209-015-0402-7

Liu, R., Zhen, R., Ding, Y., Liu, Y., Wang, J., Jiang, R., & Xu, L. (2018). Teacher support and math engagement: Roles of academic self-efficacy and positive emotions. *Educational Psychology*, *38*(1), 3–16. https://doi.org/10.1080/01443410.2017.1359238

Lodewyk, K. R. (2018). Associations between trait personality, anxiety, self-efficacy and intentions to exercise by gender in high school physical education. *Educational Psychology*, *38*(4), 487–501. https://doi.org/10.1080/01443410.2017.1375081

Martin, A. J., & Lazendic, G. (2018). Achievement in large-scale national numeracy assessment: An ecological study of motivation and student, home, and school predictors. *Journal of Educational Psychology*, *110*(4), 465–482. http://dx.doi.org/10.1037/edu0000231

Masrek, M. N., & Gaskin, J. E. (2016). Assessing users satisfaction with web digital library: The case of Universiti Teknologi MARA. *The International Journal of Information and Learning Technology*, *33*(1), 36–56. http://dx.doi.org/10.1108/IJILT-06-2015-0019

Mok, M. M. C., Chin, M. K., Chen, S., Emeljanovas, A., Mieziene, B., Bronikowski, M., Laudanska-Krzeminska, I., Milanovic, I., Pasic, M., Balasekaran, G., Phua, K. W., & Makaza, D. (2015). Psychometric properties of the attitudes toward physical activity scale: A Rasch analysis based on data from five locations. *Journal of Applied Measurement*, *16*(4), 379–400.

Mok, M. M. C., Zhu, J., & Law, C. L. K. (2017). Crosslagged cross-subject bidirectional predictions among achievements in mathematics, English language and Chinese language of school children. *Educational Psychology*, *37*(10), 1259–1280. https://doi.org/10.1080/01443410.2017.1334875

Onn, C. Y., Yunus, J. N. B., Yusof, H. B., Moorthy, K., & Na, S. A. (2018). The mediating effect of trust on the dimensionality of organizational justice and organisational citizenship behaviour amongst teachers in Malaysia. *Educational Psychology*. https://doi.org/10.1080/01443410.2018.1426836

Phan, H. P., Ngu, B. H., & Alrashidi, O. (2018). Contextualised self-beliefs in totality: An integrated framework from a longitudinal perspective. *Educational Psychology*, *38*(4), 411–434. https://doi.org/10.1080/01443410.2017.1356446

Poong, Y. S., Yamaguchi, S., & Takada, J. (2017). Investigating the drivers of mobile learning acceptance among young adults in the World Heritage town of Luang Prabang, Laos. *Information Development*, *33*(1), 57–71. https://doi.org/10.1177/0266666916638136

Sánchez-Prieto, J. C., Olmos-Miguelánez, S., & García-Peñalvo, F. J. (2017). MLearning and pre-service teachers: An assessment of the behavioral intention using an expanded TAM model. *Computers in Human Behavior*, *72*, 644–654.

Scherer, R., Rohatgi, A., & Hatlevik, O. E. (2017). Students' profiles of ICT use: Identification, determinants, and relations to achievement in a computer and information literacy test. *Computers in Human Behavior*, *70*, 486–499.

Sobel, M. E. (1982). Asymptotic confidence intervals for indirect effects in structural equation models. *Sociological Methodology*, *13*, 290–312. https://doi.org/10.2307/270723

Teo, T., Sang, G., Mei, B., & Hoi, C. K. W. (2018). Investigating pre-service teachers' acceptance of Web 2.0 technologies in their future teaching: A Chinese perspective. *Interactive Learning Environments*. https://doi.org/10.1080/10494820.2018.1489290.

Teo, T., & Zhou, M. (2017). The influence of teachers' conceptions of teaching and learning on their technology acceptance. *Interactive Learning Environments, 25*(4), 513–527. https://doi.org/10.1080/10494820.2016.1143844

Wu, J., Tennyson, R. D., & Hsia, T. L. (2010). A study of student satisfaction in a blended e-learning system environment. *Computer & Education, 55*(1), 155–164.

Yim, J. S. C., Moses, P., & Azalea, A. (2019). Predicting teachers' continuance in a virtual learning environment with psychological ownership and the TAM: A perspective from Malaysia. *Educational Technology Research and Development, 67*(3), 691–709. https://doi.org/10.1007/s11423-019-09661-8

Yuan, S., Liu, Y., Yao, R., & Liu, J. (2016). An investigation of users' continuance intention towards mobile banking in China. *Information Development, 32*(1), 20–34. https://doi.org/10.1177/0266666914522140

Zhu, M., Urhahne, D., & Rubie-Davies, C. M. (2018). The longitudinal effects of teacher judgement and different teacher treatment on students' academic outcomes. *Educational Psychology, 38*(5), 648–668. https://doi.org/10.1080/01443410.2017.1412399

Appendix

The Survey Questionnaire (English version)

General Guidelines

This questionnaire attempts to investigate "**Assessment of ICT in Higher Education Applying the TSM**"

Direction

The questionnaire has five sections. For all multiple-choice questions, please indicate your response by placing a tick (/) in the appropriate box. For seven-point Likert scale questions, please circle the number of your choice. If you wish to comment on any question or qualify your answer, please feel free to use the space in the margin or write your comments on a separate sheet of paper. All information that you provide will be kept strictly confidential. Thanks for your gracious cooperation.

Section I: Demographic information

1. Gender: Male ☐
            Female ☐

2. Age: 25–29 years old ☐
         30–34 years old ☐
         35–39 years old ☐
         40–44 years old ☐
         45 years and above ☐

3. Number of years working at the university

    1. 1–5 years
    2. 6–10 years
    3. 11–15 years
    4. 16–20 years
    5. 21 years and above

4. Highest level of educational background

    a. Undergraduate
    b. Master
    c. PhD

5. Designation:

    a. Professor
    b. Associate Professor
    c. Assistant Professor/Senior Lecturer
    d. Lecturer

6. Faculty/School_____
7. Nationality_____

Section II: Perceived ease of use of ICT facilities

For questions 1–15, rate how much you agree with each statement using the following scale:

1 = Strongly Disagree (SD)   4 = Neither agree nor disagree (N)
2 = Moderately Disagree (MD)   5 = Slightly Agree (SLA)
3 = Slightly Disagree (SLD)   6 = Moderately Agree (MA)
7 = Strongly Agree (SA)

No.	Items	SD	MD	SLD	N	SLA	MA	SA
1	I find the university ICT facilities easy to use.	1	2	3	4	5	6	7
2	I find it easy to access the university research databases.	1	2	3	4	5	6	7
3	It is easy for me to become skillful at using university databases for conducting research.	1	2	3	4	5	6	7
4	Interacting with the research databases system requires minimal effort on my part.	1	2	3	4	5	6	7
5	I find it easy to get the research databases to help facilitate my research.	1	2	3	4	5	6	7
6	Interacting with the university research databases system is very stimulating for me.	1	2	3	4	5	6	7
7	I find it easy to select articles/journals of different categories using research databases (e.g., education, engineering and business).	1	2	3	4	5	6	7
8	With Wireless Internet, I find it easy to access university databases to do research.	1	2	3	4	5	6	7

(*Continued*)

30 *Applying the Rasch Model and Structural Equation Modeling to Higher Education*

(Continued)

No.	Items	SD	MD	SLD	N	SLA	MA	SA
9	Wireless Internet allows me to access research and learning materials from Web Browser (WWW).	1	2	3	4	5	6	7
10	Wireless Internet is easy to use for teaching and research.	1	2	3	4	5	6	7
11	I find it easy to use computers provided by the university.	1	2	3	4	5	6	7
12	My interaction with university ICT services available is clear and understandable.	1	2	3	4	5	6	7
13	I find it easy to download teaching, learning and research materials using Wireless Internet in terms of speed.	1	2	3	4	5	6	7
14	It is easy for me to use multimedia facilities at the university.	1	2	3	4	5	6	7
15	I find it easy to use Microsoft Office for teaching and research purposes.	1	2	3	4	5	6	7

Section III: Perceived usefulness of ICT facilities

For questions 1–16, rate how much you agree with each statement using the following scale:

1 = Strongly Disagree (SD) 4 = Neither agree nor disagree (N)
2 = Moderately Disagree (MD) 5 = Slightly Agree (SLA)
3 = Slightly Disagree (SLD) 6 = Moderately Agree (MA)
7 = Strongly Agree (SA)

No.	Items	SD	MD	SLD	N	SLA	MA	SA
1	Using the ICT facilities at the university enables me to accomplish tasks more quickly.	1	2	3	4	5	6	7
2	Using the ICT services available at the university increases my research productivity.	1	2	3	4	5	6	7
3	The current university ICT system makes work more interesting.	1	2	3	4	5	6	7
4	ICT facilities at the university improve the quality of the work I do.	1	2	3	4	5	6	7
5	Using the university ICT facilities enhances my research skills.	1	2	3	4	5	6	7
6	Using the university ICT facilities would make it easier for me to find information.	1	2	3	4	5	6	7
7	ICT services at the university make information always available to users.	1	2	3	4	5	6	7
8	Using the ICT facilities helps me write my journal articles.	1	2	3	4	5	6	7

No.	Items	SD	MD	SLD	N	SLA	MA	SA
9	My job would be difficult to perform without the ICT facilities.	1	2	3	4	5	6	7
10	Using the ICT facilities saves my time.	1	2	3	4	5	6	7
11	Using the university ICT facilities provides me with the latest information on specific areas of research.	1	2	3	4	5	6	7
12	I can use ICT facilities from anywhere, anytime at the campus.	1	2	3	4	5	6	7
13	I believe that the use of ICT in the classroom enhances student learning in my discipline.	1	2	3	4	5	6	7
14	I believe that e-mail and other forms of electronic communication are important tools in faculty/student communication.	1	2	3	4	5	6	7
15	I believe that web-based instructional materials enhance student learning.	1	2	3	4	5	6	7
16	Using the ICT services enables me to download teaching, learning and research materials from the internet.	1	2	3	4	5	6	7

Section IV: Computer self-efficacy of ICT facilities

For questions 1–15, rate how much you agree with each statement using the following scale:

1 = Strongly Disagree (SD)   4 = Neither agree nor disagree (N)
2 = Moderately Disagree (MD)   5 = Slightly Agree (SLA)
3 = Slightly Disagree (SLD)   6 = Moderately Agree (MA)
7 = Strongly Agree (SA)

No.	Items	SD	MD	SLD	N	SLA	MA	SA
1	I have the skills required to use computer applications for writing my research papers.	1	2	3	4	5	6	7
2	I have the skills and knowledge required to use computer applications for demonstrating specific concepts in class.	1	2	3	4	5	6	7
3	I have the skills required to use computer applications for presenting lectures.	1	2	3	4	5	6	7
4	I have the skills required to communicate electronically with my colleagues and students.	1	2	3	4	5	6	7
5	I have the ability to e-mail, chat, download teaching, learning and research materials, and search different websites.	1	2	3	4	5	6	7

(Continued)

32   *Applying the Rasch Model and Structural Equation Modeling to Higher Education*

(Continued)

No.	Items	SD	MD	SLD	N	SLA	MA	SA
6	I have the ability to use the Wireless Internet service provided by the university.	1	2	3	4	5	6	7
7	I have the skills required to use ICT facilities to enhance the effectiveness of my teaching, learning and research.	1	2	3	4	5	6	7
8	I feel capable of using the university research databases for writing journal papers.	1	2	3	4	5	6	7
9	I have the ability to navigate my way through the ICT facilities.	1	2	3	4	5	6	7
10	I have the skills required to use the ICT facilities to enhance the quality of my research works.	1	2	3	4	5	6	7
11	I have the ability to save and print journals/articles from the research databases.	1	2	3	4	5	6	7
12	I have the knowledge and skills required to benefit from using the university ICT facilities.	1	2	3	4	5	6	7
13	I am capable of accessing the research databases from the university website.	1	2	3	4	5	6	7
14	I am capable of downloading and installing research-related software using ICT facilities at the university.	1	2	3	4	5	6	7
15	I am capable of using multimedia facilities for delivering lecturers.	1	2	3	4	5	6	7

Section V: Satisfaction of ICT Facilities

For questions 1–10, rate how much you agree with each statement using the following scale:

1 = Very Unsatisfied (VU)   4 = Not Sure (NS)
2 = Moderately Unsatisfied (MU) 5 = Slightly Satisfied (SLS)
3 = Slightly Unsatisfied (SLU)   6 = Moderately Satisfied (MS)
7 = Very Satisfied (VS)

No.	Items	VU	MU	SLU	NS	SLS	MS	VS
1	Overall, I am satisfied with the ease of completing my task using ICT facilities.	1	2	3	4	5	6	7
2	I am satisfied with the ICT facilities provided by the university.	1	2	3	4	5	6	7
3	I am satisfied with the ease of use of the ICT facilities.	1	2	3	4	5	6	7

No.	Items	VU	MU	SLU	NS	SLS	MS	VS
4	The university ICT facilities have greatly affected the way I search for information and conduct my research.	1	2	3	4	5	6	7
5	Providing the ICT is an indispensable service provided by the university.	1	2	3	4	5	6	7
6	Overall, I am satisfied with the amount of time it takes to complete my task.	1	2	3	4	5	6	7
7	I am satisfied with the structure of accessible information (available as categories of research domain, or by date of issue – of journals in particular, or as full-texts or abstracts of theses and dissertations) of the university research databases.	1	2	3	4	5	6	7
8	I am satisfied with the support information provided by the university's ICT facilities.	1	2	3	4	5	6	7
9	I am satisfied in using ICT facilities for teaching, learning and research.	1	2	3	4	5	6	7
10	Overall, I am satisfied with the Wireless Internet service provided at the university.	1	2	3	4	5	6	7

34 *Applying the Rasch Model and Structural Equation Modeling to Higher Education*
(Chinese Version)

使用指南

该问卷旨在调查"ICT 在高等教育的评估 – 基于技术满意度模型的应用"

说明

该问卷由5个小部分组成。衡量方式采用李克特七点尺度量表，请在每一道题后面的**选项**中用(√)标出您的选择。如果您对任何问题有不同的看法或有不同的答案，您可以写在对应题号左右的空白处，或写在另外纸上的空白处。您所填的个人信息将会被严格保密。谢谢您的合作。

第一部分：个人信息

1. 性别：男 □ 女 □

2. 年龄：25–29 岁 □
   30–34 岁 □
   35–39 岁 □
   40–44 岁 □
   45岁及以上 □

3. 大学的工作经历：
   a. 1–5 年
   b. 6–10 年
   c. 11–15 年
   d. 16–20 年
   e. 21年及以上

4. 本人已获得的最高学历：
   a. 本科
   b. 硕士
   c. 博士

5. 职称：
   a. 教授
   b. 副教授
   c. 助理教授
   d. 高级讲师
   e. 讲师

第二部分：使用信息和通信设备的易用性感知

问题1–15，用以下几种标准来评定你对每一项的认同程度，请用(√)标出您的选择。

1 = 强烈不赞同   4 = 既不赞同也不反对
2 = 中度不赞同   5 = 稍赞同
3 = 略不赞同   6 = 中度赞同
7 = 强烈赞同

序号	陈述项目	强烈不赞同	中度不赞同	略不赞同	既不赞同也不反对	稍赞同	中度赞同	强烈赞同
1	我觉得学校的信息和通信设备容易使用。	1	2	3	4	5	6	7
2	我觉得访问学校的研究性数据库是容易的。	1	2	3	4	5	6	7

序号	陈述项目	强烈不赞同	中度不赞同	略不赞同	既不赞同也不反对	稍赞同	中度赞同	强烈赞同
3	对我来说，我能容易的就变得熟练地使用学校数据库来做研究。	1	2	3	4	5	6	7
4	与研究性数据库系统互动需要我很少的努力。	1	2	3	4	5	6	7
5	我觉得使用数据库来帮助促进我的研究是容易的。	1	2	3	4	5	6	7
6	对我来说，与学校研究性数据库进行互动很令人兴奋。	1	2	3	4	5	6	7
7	我觉得使用研究性数据库来选择不同类别的文章/期刊是容易的（比如：教育类/工程类/商务类/语言类等）。	1	2	3	4	5	6	7
8	我觉得使用无线网络访问学校数据库做研究是容易的。	1	2	3	4	5	6	7
9	无线网络允许我通过网页浏览器（WWW万维网）访问研究和学习资源。	1	2	3	4	5	6	7
10	使用无线网络来进行教学和研究是容易的。	1	2	3	4	5	6	7
11	我觉得学校提供的计算机使用起来是容易的。	1	2	3	4	5	6	7
12	对我来说，与学校可获得的信息和通信服务的互动过程是明确的、易懂的。	1	2	3	4	5	6	7
13	在网络方面，我觉得使用无线网络下载教学、学习和研究资料是容易的。	1	2	3	4	5	6	7
14	对我来说，在学校使用多媒体设施是容易的或方便的。	1	2	3	4	5	6	7
15	我觉得微软办公应用软件（Microsoft Office）在以教学和研究为目的的应用中是容易的。	1	2	3	4	5	6	7

第三部分：信息和通信设备的实用性感知

问题1–16，用以下几种标准来评定你对每一项的认同程度，请用(√)标出您的选择。

1 = 强烈不赞同　　　4 = 既不赞同也不反对
2 = 中度不赞同　　　5 = 稍赞同
3 = 略不赞同　　　　6 = 中度赞同
7 = 强烈赞同

序号	陈述项目	强烈不赞同	中度不赞同	略不赞同	既不赞同也不反对	稍赞同	中度赞同	强烈赞同
1	使用学校的信息和通信设施能够使我更快地完成任务。	1	2	3	4	5	6	7
2	使用学校可获得的信息和通信服务提高我的研究生产力。	1	2	3	4	5	6	7
3	当前学校的信息与通信系统使工作变得更加有趣。	1	2	3	4	5	6	7
4	学校的信息与通信设施提高我的工作质量。	1	2	3	4	5	6	7
5	使用学校的信息与通信设施增强我的研究技能。	1	2	3	4	5	6	7
6	对我来说，使用学校的信息与通信设施寻找信息会更容易。	1	2	3	4	5	6	7
7	学校的信息与通信服务让用户始终可获得信息。	1	2	3	4	5	6	7
8	使用信息与通信设施有助于我写期刊文章。	1	2	3	4	5	6	7
9	没有信息与通信设施，我的工作会难以执行。	1	2	3	4	5	6	7
10	使用信息与通信设施节省我的时间。	1	2	3	4	5	6	7
11	使用学校的信息与通信设施给我提供特定研究领域的最新信息。	1	2	3	4	5	6	7
12	我可以在校园里的任意时间和任何地方使用信息与通信设施。	1	2	3	4	5	6	7

序号	陈述项目	强烈不赞同	中度不赞同	略不赞同	既不赞同也不反对	稍赞同	中度赞同	强烈赞同
13	在我的课上，我认为信息与通信设施的使用提高学生的学习能力。	1	2	3	4	5	6	7
14	我认为电子邮件和其他形式的电子通讯是老师/学生交流沟通的重要工具。	1	2	3	4	5	6	7
15	我认为基于网络的教学材料提高学生的学习能力。	1	2	3	4	5	6	7
16	使用信息与通信服务允许我从英特网上下载教学，学习和研究资料。	1	2	3	4	5	6	7

第四部分：自我效能

问题1–15，用以下几种标准来评定你对每一项的认同程度，请用(√)标出您的选择。

1 = 强烈不赞同   4 = 既不赞同也不反对
2 = 中度不赞同   5 = 稍赞同
3 = 略不赞同   6 = 中度赞同
7 = 强烈赞同

序号	陈述项目	强烈不赞同	中度不赞同	略不赞同	既不赞同也不反对	稍赞同	中度赞同	强烈赞同
1	我拥有所需的技能使用相关计算机应用来写我的研究论文。	1	2	3	4	5	6	7
2	我拥有所需的技能和知识使用相关计算机应用来在课堂上展现出特定具体的概念。	1	2	3	4	5	6	7
3	我拥有所需的技能使用相关计算机应用来授课。	1	2	3	4	5	6	7
4	我拥有所需的技能来与我的同事和学生进行电子通讯交流。	1	2	3	4	5	6	7
5	我有能力发邮件，聊天，下载教学、学习和研究资料，以及搜索不同的网站。	1	2	3	4	5	6	7

(Continued)

38 *Applying the Rasch Model and Structural Equation Modeling to Higher Education*

(Continued)

序号	陈述项目	强烈不赞同	中度不赞同	略不赞同	既不赞同也不反对	稍赞同	中度赞同	强烈赞同
6	我有能力使用学校提供的无线网络服务。	1	2	3	4	5	6	7
7	我拥有所需的技能使用信息与通信设备来提高我的教学、学习与研究效益。	1	2	3	4	5	6	7
8	我相信我有能力使用学校的研究性数据库来写期刊论文。	1	2	3	4	5	6	7
9	我有能力通过我的方式来使用信息与通信设施。	1	2	3	4	5	6	7
10	我拥有所需的技能使用信息与通信设施来提高我研究工作的质量。	1	2	3	4	5	6	7
11	我有能力从研究性数据库中保存和打印期刊/文章。	1	2	3	4	5	6	7
12	我拥有所需的知识和技能从学校信息与通信设施的使用中获益。	1	2	3	4	5	6	7
13	我能够从学校网站上访问研究性数据库资源。	1	2	3	4	5	6	7
14	我能够使用学校的信息与通信设施下载和安装与研究相关的软件。	1	2	3	4	5	6	7
15	我能够使用多媒体设施来授课。	1	2	3	4	5	6	7

第五部分：对信息和通信设备的满意度

问题1–10，用以下几种标准来评定你对每一项的认同程度，请用(√)标出您的选择。

1 = 非常不满意　　4 = 不确定
2 = 中度不满意　　5 = 稍满意
3 = 略不满意　　　6 = 中度满意
7 = 非常满意

序号	陈述项目	非常不满意	中度不满意	略不满意	不确定	稍满意	中度满意	非常满意
1	总体上，我对使用信息与通信设施完成任务的简易性表示满意。	1	2	3	4	5	6	7
2	我满意于学校提供的信息与通信设施。	1	2	3	4	5	6	7
3	我满意于使用信息与通信设施的简易性。	1	2	3	4	5	6	7
4	学校的信息与通信设施已经很大程度上影响了我搜索信息和开展研究的方式。	1	2	3	4	5	6	7
5	信息与通信服务是学校提供的必不可少的服务。	1	2	3	4	5	6	7
6	总体上，我满意于完成任务所花的时间量。	1	2	3	4	5	6	7
7	我满意于学校数据库可访问信息的布局构造（以研究领域分类，或以特定期刊、论文全文或摘要的发表日期呈现）。	1	2	3	4	5	6	7
8	我满意于学校的信息与通信设施所提供的支持信息。	1	2	3	4	5	6	7
9	我满意于使用信息与通信设施来教学，学习和做研究。	1	2	3	4	5	6	7
10	总体上，我满意于学校所提供的无线网络服务。	1	2	3	4	5	6	7

2 Testing Online Learning Satisfaction in Higher Education

2.1 Introduction

In an effort to mitigate the spread of the COVID-19 pandemic, most educational institutions around the world have been closed since about March 2020. This has impacted more than 90% of the world's student population (UNESCO, 2020a). In China, educational institutions were closed in late January 2020 (Zhu & Peng, 2020). The ministry of education of China (2020a) initiated an emergency policy called "Suspending Classes Without Stopping Learning" to make sure that students could continue their studies with online learning platforms at home on January 29. In the field of higher education, between January 29 and April 3, 1,454 Chinese universities started the spring semester using online learning platforms nationwide. Over 950,000 university teachers offered more than 942,000 courses and 7,133,000 lectures on online learning platforms. In total, university students have attended these courses and lectures 1.18 billion times (Ministry of Education, 2020b). This online learning practice in China is unprecedented in scale, scope and depth and is considered the first exploration of its kind in the history of higher education worldwide. Moreover, helping students adapt to a new learning pattern fully integrated with information and communication technologies (ICTs) has been a very important experiment (Ministry of Education, 2020b).

In recent years, an increasing number of technology platforms have been widely adopted to support learning in higher education (de Souza Rodrigues et al., 2021; Habib et al., 2021; Mpungose, 2020; Su & Chen, 2020; Turnbull et al., 2019; Yen et al., 2018; Yunusa & Umar, 2021). For instance, learning management systems (LMSs) have been considered to be one of the most important and indispensable online learning tools and platforms (Coates et al., 2005; Turnbull et al., 2021). LMSs can be defined as "web-based software platforms that provide an interactive online learning environment and automate the administration, organization, delivery, and reporting of educational content and learner outcomes" (Turnbull et al., 2019, p. 1). LMSs have many features that support online learning, including course management, assessment, learner progress tracking, gradebook, communications, security and smartphone access (Turnbull et al., 2019, 2021). These features can work together to provide a

DOI: 10.1201/9781003384724-2

seamless experience for online learners (Turnbull et al., 2019). The quality of online learning is also influenced by the robustness of learning platforms (Pinho et al., 2021; Uppal et al., 2018). Furthermore, these platforms have been recognised as irreplaceable emergency educational tools in the transition to online learning during the COVID-19 pandemic (Zhu & Peng, 2020).

The success of online learning platforms (e.g., LMSs) has generally been determined by student satisfaction (Virtanen et al., 2017; Yuen et al., 2019). Dai et al. (2020) found evidence of a relationship between higher student satisfaction and more positive attitudes toward LMSs. Other studies found that student satisfaction was related to stronger intention and willingness to use (Salam & Farooq, 2020), higher long-term adoption rate (Cidral et al., 2018), better learning performance (Al-Fraihat et al., 2020; Isaac et al., 2019) and achievements (Vasileva-Stojanovska et al., 2015). Moreover, student satisfaction is also a powerful influential factor involved in a platform's or a system's net benefits (Martins et al., 2018; Salam & Farooq, 2020). Although the governments, universities and service providers have made significant investments in new technologies, the full benefit and value of online learning platforms have not yet been realised (Barclay et al., 2018; Lane et al., 2015), nor have students yet been as satisfied as expected (Chingos et al., 2017; Deng et al., 2019; Jiang & Zhao, 2018). This necessitates the implementation of continuous investigations on determinants of student satisfaction (Fırat et al., 2018; Herrador-Alcaide et al., 2019). On the one hand, governments, universities and service providers can target areas that need to be changed and improved based on the determinants of student satisfaction and thus enhance the online learning service quality through scientific, appropriate, effective and reliable methods (Cidral et al., 2018; Machado-Da-Silva et al., 2014). On the other hand, educators, course developers and instructional designers can also benefit from these investigations and thus provide students with better online learning environments and more suitable online learning programmes (Cidral et al., 2018; Ilgaz & Gülbahar, 2015). However, prior studies on online learning were conducted mostly in developed countries, and limited effort has been made in emerging countries (Pham et al., 2019).

Motivated by these gaps, the main aim of this study was to validate the technology satisfaction model (TSM) in order to explore the determinants of university students' satisfaction with using online learning platforms in the context of Chinese higher education during the COVID-19 pandemic. Subsequently, this study validated the TSM among Eastern and Western Chinese university students and scrutinised regional differences. Governments, universities, platform service providers, educators, course developers and instructional designers can use the findings as a basis to improve the service quality of online learning, enhance university students' satisfaction and increase contingency capacities in order to mitigate and manage risk in the future.

The remainder of the paper is organised as follows. In the next section, the background of the study is introduced, after which state-of-the-art related studies are presented. After that, the research method adopted for this study

42   *Applying the Rasch Model and Structural Equation Modeling to Higher Education*

is outlined. Results follow in the subsequent section. The discussion section focuses on the contributions of this study and the theoretical and practical implications, and finally, the conclusion and limitations are outlined.

2.2 Background

Online learning platforms have played an irreplaceable role in the massive practice of online learning in China during the COVID-19 pandemic. In the early days of school closures, university students all over China faced the dilemma of "waiting to learn at home" (Zhu & Peng, 2020, p. 1). In order to deal with the dilemma swiftly and respond to the public's concerns, the Ministry of Education (2020c, 2020d) issued a series of emergency measures which included organising multiple platforms to support university students' online learning. By April 3, there were 37 government-backed platforms (e.g., XuetangX, Eduyun, etc.) and more than 110 social and university platforms (e.g., Daxiaxuetang, Cqooc, etc.) across the country involved in providing university students with online learning resources and services (Ministry of Education, 2020b). Different students were required to take different courses on different platforms to avoid excessive pressure on the servers. Fortunately, these platforms could generally meet the massive online learning demands of more than 31.04 million university students in China. In view of this, more and more Chinese administrators and scholars began to shift their focus from whether students could learn to whether they could learn well and be satisfied with their learning environments (Zhu et al., 2020).

Since serving students and ensuring their satisfaction are the fundamental goals of promoting e-education in China, the Ministry of Education (2019) has launched a series of initiatives in recent years to develop and improve online learning platforms. However, students in technology-enhanced sessions have reported significantly lower satisfaction than those in traditional classrooms (Chingos et al., 2017; Deng et al., 2019), and Chinese students are no exception (Jiang & Zhao, 2018). Some studies have investigated Chinese university students' intention to use learning management systems or virtual and remote labs (Su & Chen, 2020; Zhang et al., 2020). However, as far as we know, few studies in China have explored the determinants of university students' satisfaction with using online learning platforms, let alone during the COVID-19 pandemic. As a consequence, governments, universities and platform service providers may have little strategic guidance to enhance students' satisfaction.

Another concern of Chinese administrators and scholars is whether online learning will exacerbate educational inequality (Hu & Xie, 2020; Yang, 2020). Previous studies in China have shown that online learning platforms expose more students to rich educational resources but do not benefit all social classes equally; disadvantaged student groups (e.g., rural students, students with low socioeconomic status) frequently benefit less from them (Xu & Yao, 2018; Xu & Ye, 2018). During the COVID-19 pandemic, these issues regarding educational equity were raised again (Kingsbury, 2021; UNESCO, 2020b). There is a huge development gap (e.g., human resources gap, financial resources gap and material resources gap, etc.) between Eastern and Western Chinese universities (Cai et al., 2021). For instance, in 2017, Western Chinese universities' overall financial education funds were allocated about 240,763 million yuan, while Eastern Chinese universities were allocated about 612,898 million yuan (Cai et al., 2021). The latest assessment in China also illustrates the gap in development, as Western Chinese universities have 51 first-class disciplines in total while Eastern Chinese universities have 331 (Cai et al., 2021). In view of this, scrutinising the difference between Eastern and Western Chinese university students' satisfaction with platforms may help us better understand online learning equity. However, as far as we know, very few studies have discussed these equity issues from the perspective of regional differences in developing countries.

2.3 Literature Review

According to Islam et al. (2018), the present study defines student satisfaction as the degree to which "the use of technology is consistent with existing values, needs and student experiences" in the use of online learning platforms (Islam et al., 2018, p. 4). The technology satisfaction model is one of the most important models that has been validated as effective in explaining students' satisfaction in Asian higher educational settings. Proposed by Islam (2014), this model combines two psychological factors, namely satisfaction and computer self-efficacy (from Bandura's (1977) social cognitive theory), with two motivation variables, namely perceived ease of use and usefulness (from Davis et al.'s (1989) technology acceptance model (TAM)). TAM has its foundation in the theory of reasoned action (TRA), a general theory widely applied to predict and explain human behaviour in a variety of contexts (Ajzen & Fishbein, 1980). In educational contexts, TAM has become one of the most important and popular models in understanding predictors of teachers' and students' ICT acceptance (Granić & Marangunić, 2019; Scherer et al., 2019). It asserts that perceived ease of use and usefulness are two main determinants of an individual's intention to use and attitude toward using technology (Davis et al., 1989). Despite its broad applicability and strong explanatory power, some critics have claimed that it focuses on behavioural and motivational factors while ignoring psychological factors, such as computer self-efficacy (Scherer & Teo, 2019; Yalçın & Kutlu, 2019) and satisfaction (Scherer & Teo, 2019; Yuen et al., 2019). The original TSM (Islam, 2014) took these two essential psychological factors into consideration and articulated three influential determinants of technology satisfaction (SAT), namely computer self-efficacy (CSE), perceived ease of use (PEU) and perceived usefulness (PU) (see Figure 2.1). The TSM was validated in a Malaysian university to measure students' satisfaction with using online research databases (Islam et al., 2015) and wireless internet (Islam, 2014). It was later validated in a Chinese university to measure students' satisfaction with using wireless internet in learning activities (Islam et al., 2018). In 2020, Islam

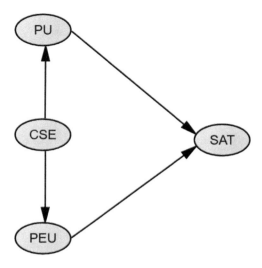

Figure 2.1 Technology satisfaction model (Islam, 2014).

Note: satisfaction (SAT), computer self-efficacy (CSE), perceived ease of use (PEU) and perceived usefulness (PU).

and Sheikh (2020) validated the relationships within the TSM in a Pakistani university. However, the TSM has not been validated to assess online learning platform success. Nevertheless, more universities in different regions should be included to enhance the TSM's applicability (Islam et al., 2015).

2.4 Hypotheses

In this section, this study presents a brief but pertinent review of theoretical and empirical literature regarding the determinants of university students' satisfaction with using online learning platforms. Based on the TSM, this study devised a total of seven hypotheses and thus clarified the relationships among the four latent variables of the model.

Self-efficacy, as an important component of Bandura's (1977) social cognitive theory, was defined as "people's beliefs about their capabilities to produce designated levels of performance that exercise influence over events that affect their lives" (Bandura, 1997, p. 71). Based on Bandura's (1977, 1997) theory, Venkatesh and Davis (1996) adapted the concept of computer self-efficacy to people's judgment of their capabilities to easily use information and computer technologies. Regarding educational technology, Islam et al. (2015, p. 57) referred to computer self-efficacy as "student's beliefs in their capabilities to use a computer for their learning and research". Following this definition, this study posits that computer self-efficacy is associated with university students' beliefs in their capabilities to use online learning platforms for their study. Extensive studies have shown that computer self-efficacy significantly impacts learners in various ways within a technology-supported environment (Dong et al., 2020; Heckel & Ringeisen, 2019; Wang et al., 2019; Zhu & Mok, 2020). Most importantly, recent studies have found that computer self-efficacy is directly related to the two key motivation variables in the TAM, perceived ease of use and usefulness (Bin et al., 2020; Chen et al., 2019; Scherer et al., 2019; Thongsri et al., 2019; Yalçın & Kutlu, 2019). In other words, it is probable that university students with high computer self-efficacy find it easy to use online learning platforms as well as realise their value and benefits. In view of this, Islam and Sheikh (2020) suggested that more attention should be paid to the assessment of computer self-efficacy and its impact on students' perceived ease of use and usefulness. Therefore, this study hypothesises the following:

H1: Eastern and Western Chinese students' computer self-efficacy directly impacts the perceived ease of use of online learning platforms.
H2: Eastern and Western Chinese students' computer self-efficacy directly impacts the perceived usefulness of online learning platforms.

Drawing from Davis et al.'s (1989) and Islam's (2011) definitions, the current study refers to perceived usefulness as university students' perception of the benefits of using online learning platforms, and it refers to perceived ease of use as university students' perception of how easy or difficult it is to use online learning platforms. On the one hand, perceived ease of use and usefulness have frequently been considered core variables in different studies on online learning (Ameen et al., 2019; Esteban-Millat et al., 2018; Farhan et al., 2019; Scherer et al., 2019). On the other hand, technology satisfaction is also directly impacted by perceived ease of use and usefulness (Bin et al., 2020; Chen et al., 2019; Islam, 2016; Islam et al., 2018; Islam & Sheikh, 2020). However, to our knowledge, students' satisfaction with using online learning platforms has seldom been measured by perceived ease of use and usefulness in Chinese higher education. Thus, we have proposed the following hypotheses:

H3: Perceived ease of use directly impacts Eastern and Western Chinese students' satisfaction with using online learning platforms.
H4: Perceived usefulness directly impacts Eastern and Western Chinese students' satisfaction with using online learning platforms.

Many previous studies have validated the significant associations between computer self-efficacy, perceived ease of use and usefulness (Abdullah & Ward, 2016; Bin et al., 2020; Chen et al., 2019; Scherer et al., 2019; Thongsri et al., 2019; Yalçın & Kutlu, 2019). For instance, according to Abdullah and Ward's (2016) meta-analysis, computer self-efficacy is the greatest predictor of students' perceptions of the ease of use of online learning systems or platforms, and it is also an important predictor of how useful students perceive them to be. Moreover, Chen et al. (2019), Islam et al. (2018), and Islam and Sheikh (2020) argued that learners' satisfaction can also be influenced by computer self-efficacy and mediated by perceived ease of use and usefulness. However, very few studies on online learning platforms have validated the indirect impact in China. Thus, this study hypothesises:

> *H5*: Eastern and Western Chinese students' computer self-efficacy indirectly impacts their satisfaction mediated by the perceived usefulness of online learning platforms.
>
> *H6*: Eastern and Western Chinese students' computer self-efficacy indirectly impacts their satisfaction mediated by the perceived ease of use of online learning platforms.

It is worth noting that some of the relationships and influences mentioned earlier may be moderated by cultural or regional factors (Huang, 2017; Islam, 2016). This has motivated some researchers to do cross-cultural or cross-regional analyses when assessing ICT acceptance, adoption and satisfaction (Hassan & Wood, 2020; Huang, 2017; Islam, 2016; Jung & Lee, 2020). For instance, Islam (2016) found that culture did interact with computer self-efficacy, perceived ease of use, perceived usefulness and intention to use and that it influenced both Malaysian and Chinese lecturers' adoption and satisfaction with using ICT in higher education. During the COVID-19 pandemic, a large-scale survey of 277,521 students in Hubei, China, revealed that rural students' online learning satisfaction was significantly higher than city students' (Wang et al., 2020). However, despite the huge development gap between Eastern and Western China, nearly none of the research has scrutinised the regional differences in Eastern and Western Chinese university students' satisfaction with online learning platforms. Thus, our hypothesis is as follows:

> *H7*: There will be a cross-regional invariant of the causal structure of the TSM between Eastern and Western Chinese university students.

2.5 Methodology

In line with Burkell (2003), we believed that it was best to employ a survey questionnaire method to collect Eastern and Western Chinese students' opinions, information and experiences in terms of e-learning during the COVID-19 pandemic, because administering surveys is the most popular method of data collection in studies related to online learning platforms in China (Turnbull et al., 2020). Islam (2011) designed the original questionnaire for the purpose of measuring online research databases in English. Recently, Chen et al. (2019) translated and validated it in the Chinese higher education context. In order to better suit our purpose of assessing university students' satisfaction with online learning platforms, both versions of the instrument were adapted and modified. Subsequently, the adapted instrument was pretested by giving the survey to a

Table 2.1 Dimensions of the TSM Model Measured by the Number of Items

Dimensions	Likert Scale	No. of Items
Perceived ease of use	1–6 (strongly disagree → strongly agree)	10
Perceived usefulness	1–6 (strongly disagree → strongly agree)	10
Computer self-efficacy	1–6 (strongly disagree → strongly agree)	8
Satisfaction	1–6 (very unsatisfied → very satisfied)	5
Total		33

sample of 125 university students from East China Normal University and Yuxi Normal University in order to evaluate their reaction to the items and ease of answerability. Revision was undertaken after the statistical analysis which used the Rasch model. The formal questionnaire used in this current study contained 33 items, and each item was measured by a six-point Likert scale. Table 2.1 shows the dimensions of the TSM model which the questionnaire measured. We obtained ethical endorsement for this research before distributing the questionnaires among university students.

Five universities were selected for the formal test. Among them, East China Normal University (ECNU) and Zhejiang Normal University (ZJNU) are located in eastern regions of China, while Xizang Minzu University (XZMU), Yuxi Normal University (YXNU) and Qujing Normal University (QJNU) are located in western regions. An online invitation was delivered to students of the five universities at the end of the spring semester (Semester 2, 2019–2020). During the survey, the universities' rules and regulations were followed, students' anonymity was confirmed and all personal information was strictly protected. This study recruited a total of 936 students to participate in the survey through a purposive sampling technique. According to Etikan et al. (2016), purposive sampling, which involves deliberately choosing participants based on the qualities they possess, would facilitate a focus on the regional differences of students' universities. Summarised from a preliminary analysis report, eight of the collected questionnaires were determined to be invalid due to incomplete responses. Next, data were analysed using SPSS 21.0 to conduct descriptive analysis. Winsteps software version 3.94 was used to conduct Rasch analysis for validating the instrument. AMOS software version 16 was used to perform extensive analyses for three-stage structural equation modeling like confirmatory factor analysis (CFA) and a full-fledged structural model and invariance analysis for validating the measurement and structural model and cross-validating the TSM model, respectively.

The data set consisted of 33.8% male and 66.2% female university students. Of these, 52.9% came from Eastern Chinese universities, namely ECNU and ZJNU, while 47.1% came from Western Chinese universities, namely XZMU, YXNU and QJNU. Eight per cent were 17–18 years of age, 47.5% were 19–20 years of age, 38.6% were 21–22 years of age, 5.1% were 23–24 years of age, and 0.9% were 25 years of age and older. Undergraduate students

48 *Applying the Rasch Model and Structural Equation Modeling to Higher Education*

constituted 82.8%, and 17.2% were postgraduate students. In China, 86.14% of university students are undergraduates, and 13.86% are postgraduates (National Bureau of Statistics, 2019). This ratio is close to that in our data set.

2.6 Results

Over the last 30 years, a wide range of disciplines has gradually adopted the Rasch model (Rasch, 1960) as a theory-based method for developing measurements. A clear explanation of the factors to be estimated is an essential prerequisite for developing an assessment. In order to examine the initially predicted determinant, evaluation data are often fitted to a Rasch measurement. If the data fit the model, then the assertion can be substantiated, the existence of the initially hypothetical construct can be justified, and the construct can then be measured by the psychometric properties. Accordingly, evidence for construct and content validity is proffered. Furthermore, the Rasch model also allows for various ways to test differential item functioning or item bias, which increases the possibility of developing a fair measurement scale (Liu & Boone, 2006). As a result, the present study used Winsteps software to conduct Rasch analysis. Several outputs of the Rasch analysis explained that items' reliability and their separation are quite high, i.e., .99 and 10.31, respectively. Furthermore, Rasch person reliability and its separation are highly satisfactory, i.e., 96 and 4.95, respectively. The item polarity map of the Rasch model indicated that all the scores of point measure correlation (PTMEA CORR.) for items were greater than .61 and that they measured in the same direction. However, item fit order found that three items out of 33 were outside the range of infit (> 0.5) and outfit (< 1.5) mean square (MNSQ) scores (Bond & Fox, 2001). Therefore, these three misfitting indicators (PU4, PEU6 and CSE5) were considered to be invalid, and they should be excluded from further estimation (see Appendix). According to the principal components of Rasch analysis, the remaining 30 valid items for measuring four facets empirically explained 75% of the variance and confirmed a good measurement scale as shown in Table 2.2.

Interestingly, the item map of the Rasch model (see Figure 2.2) found that the majority of Eastern and Western Chinese university students were able to respond to the items correctly, and their ability to use online learning platforms were higher than the items' difficulties. Items are located on the right-hand side of Figure 2.2 while the persons are located on the left-hand side of the figure.

Firstly, this study obtained a pool of 30 valid items through Rasch analysis using minimum likelihood estimation after which we ran through the confirmatory factor analysis using maximum likelihood estimation to validate the measurement model of our study. Basically, the measurement model was designed based on the constructs of the TSM, which were interrelated to measure the convergent and discriminant validity using a few required statistical assumptions like composite reliability (CR), average variance extracted (AVE) and covariances, including square root of AVE. Hair et al. (2010) suggested that CR and AVE scores should be larger than .70 and .50, respectively. Meanwhile,

Testing Online Learning Satisfaction in Higher Education 49

Table 2.2 The 30 Valid Items

Constructs	Valid Items		α
PEU	PEU1	I find the online learning platforms easy to use.	.945
	PEU2	I find it easy to access the online learning platforms.	
	PEU3	It is easy for me to become skillful at using the online learning platforms.	
	PEU4	It is easy for me to remember how to search and take courses by using the online learning platforms.	
	PEU5	Interacting with the online learning platforms requires minimal mental effort.	
	PEU7	I find it easy to get the online learning platforms to help me take the courses.	
	PEU8	My interaction with the online learning platforms is within my comprehension.	
	PEU9	Interacting with the online learning platforms is very stimulating for me.	
	PEU10	I find it easy to choose courses of different categories (e.g., education, engineering and business).	
PU	PU1	Using the online learning platforms enables me to get the courses I am interested in.	.956
	PU2	Using the online learning platforms helps me learn my courses.	
	PU3	The online learning platforms address my study-related needs.	
	PU5	Using the online learning platforms system saves me time.	
	PU6	Using the online learning platforms system allows me to accomplish more study work than would otherwise be possible.	
	PU7	Using the online learning platforms improves the quality of my study.	
	PU8	Using the online learning platforms enhances my knowledge and learning skills.	
	PU9	Using the online learning platforms improves my study performance.	
	PU10	Using the online learning platforms provides me with the latest information on particular areas of study.	
CSE	CSE1	I am able to use the online learning platforms.	.949
	CSE2	I can download learning materials from the online learning platforms.	
	CSE3	I can navigate my way through the online learning platforms.	
	CSE4	I can use the online learning platforms to enhance the quality of my study.	
	CSE6	I have the ability to communicate with teachers and classmates through online learning platforms.	
	CSE7	I have the knowledge and skills required to benefit from using the online learning platforms.	
	CSE8	I can access the online learning platforms from home.	
SAT	SAT1	Overall, I am satisfied with the ease of completing my task by using the online learning platforms.	.929

(*Continued*)

50 *Applying the Rasch Model and Structural Equation Modeling to Higher Education*

Table 2.2 (Continued)

Constructs	Valid Items		α
	SAT2	The online learning platforms service has greatly affected the way I learn.	
	SAT3	The online learning platforms are indispensable and satisfactory services provided for Chinese college students.	
	SAT4	Overall, I am satisfied with the amount of time it takes to complete my study task by using the online learning platforms.	
	SAT5	I am satisfied with the structure of accessible information (available as categories of courses, or by date of courses) of the online learning platforms.	

Fornell and Larcker (1981) claimed that the square root of AVEs should be larger than covariances among the facets. Our measurement model was also estimated based on several recommendations (Hu & Bentler, 1999) of fit indices, like chi-square (χ^2)/degree of freedom (< 5) including the root mean square error of approximation (RMSEA < .1), Tucker–Lewis index (TLI > .90) and comparative fit index (CFI > .90). Based on the aforementioned assumptions, four-factor measurement model of computer self-efficacy, satisfaction, perceived ease of use and perceived usefulness adjusted the data efficaciously after isolating several items due to the multicollinearity, with $\chi^2 = 443.609$; $df = 96$; $p = .000$; RMSEA = .062; CFI = .974; and TLI = .968. Moreover, the measurement model with the remaining 16 parameters confirmed the convergent and discriminant validity, where CR and AVE scores for all the factors are above 0.892 and 0.674 (see Table 2.3), including the fact that the coefficients of interrelationships among the determinants did not exceed the cut-off point of 0.85 (Fornell & Larcker, 1981), suggesting that our structure model be tested further.

In Table 2.4, this study reported 16 valid items of our measurement model and its factor loadings including Cronbach's alpha (α), mean (M) and standard deviation (SD).

The TSM is the structural model of our study. We used 16 indicators of the measurement model to obtain evidence for our hypotheses. The path diagram of the TSM shows that the structural model adjusted the data satisfactorily, with $\chi^2 = 436.654$; $df = 100$; $p = .000$; RMSEA = .076; CFI = .960; and TLI = .952 (see Figure 2.3). Six hypotheses of the TSM were tested through the path coefficients (β) and critical ratios (CRs), including p values. For instance, university students' CSE directly impacted their PEU $(\beta = .84$, CR = 22.744, $p < .000)$ and PU $(\beta = .71$, CR = 22.596, $p < .000)$ of online learning platforms and supported the first two hypotheses (e.g., *H1* and *H2*). Along this line, PEU $(\beta = .27$, CR = 9.032, $p < .000)$ and PU $(\beta = .71$, CR = 20.730, $p < .000)$

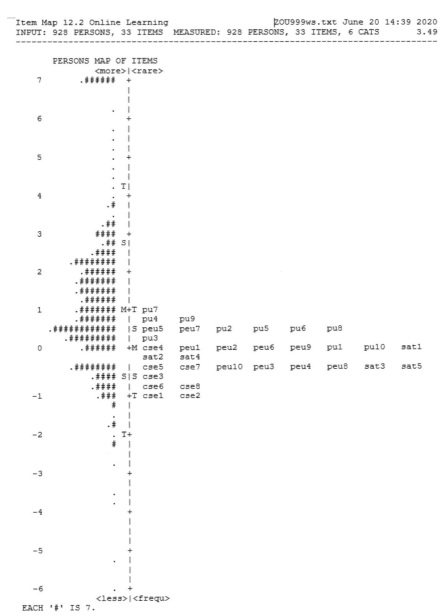

Figure 2.2 Item map.

52 *Applying the Rasch Model and Structural Equation Modeling to Higher Education*

Table 2.3 The Scores of CR, AVE and Square Root of AVE

Factors	CR	AVE	CSE	PU	SAT	PEU
CSE	0.931	0.773	**0.879**			
PU	0.916	0.731	0.679	**0.855**		
SAT	0.908	0.713	0.764	0.794	**0.844**	
PEU	0.892	0.674	0.831	0.754	0.775	**0.821**

Note: Bold numbers show the square roots of the AVEs.

Table 2.4 Valid Parameters of the Measurement Model

Factors		Measurement Variables	Loadings	M	SD	α
CSE	CSE1	I am able to use the online learning platforms.	.87	4.85	.984	.931
	CSE3	I can navigate my way through the online learning platforms.	.92	4.64	1.081	
	CSE6	I have the ability to communicate with teachers and classmates through online learning platforms.	.84	4.66	1.068	
	CSE8	I can access the online learning platforms from home.	.87	4.72	1.051	
PEU	PEU2	I find it easy to access the online learning platforms.	.80	4.37	1.145	.893
	PEU3	It is easy for me to become skillful at using the online learning platforms.	.91	4.51	1.153	
	PEU4	It is easy for me to remember how to search and take courses by using the online learning platforms.	.85	4.50	1.152	
	PEU5	Interacting with the online learning platforms requires minimal mental effort.	.74	4.09	1.233	
PU	PU2	Using the online learning platforms helps me learn my courses.	.89	4.10	1.214	.919
	PU3	The online learning platforms address my study-related needs.	.87	4.19	1.155	
	PU6	Using the online learning platforms system allows me to accomplish more study work than would otherwise be possible.	.82	4.04	1.213	
	PU8	Using the online learning platforms enhances my knowledge and learning skills.	.87	4.09	1.151	
SAT	SAT1	Overall, I am satisfied with the ease of completing my tasks by using the online learning platforms.	.90	4.40	1.134	.908
	SAT2	The online learning platform service has greatly affected the way I learn.	.75	4.30	1.113	
	SAT3	The online learning platforms are indispensable and satisfactory services provided for Chinese university students.	.85	4.46	1.082	
	SAT4	Overall, I am satisfied with the amount of time it takes to complete my study tasks by using the online learning platforms.	.85	4.29	1.176	

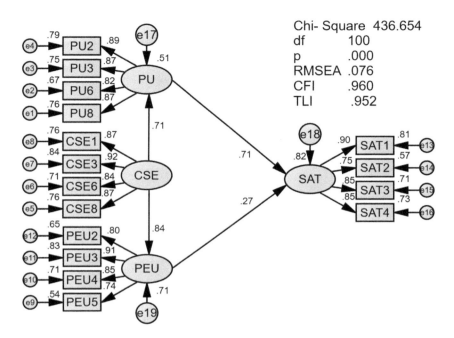

Figure 2.3 The path diagram of the TSM.

directly impacted SAT of online learning platforms, which supported our next two hypotheses (i.e., *H3* and *H4*).

This study also estimated the significant indirect impacts of the exogenous variable (CSE) on endogenous (SAT) variable through mediating variables (PEU and PU) using the Sobel test (Sobel, 1982) as our indirect hypotheses supported (e.g., *H5* and *H6*). For example, university students' CSE had an indirect impact on SAT mediated by their PEU (chi-square, χ^2 = 15.442; p = .000) and PU (chi-square, χ^2 = 8.425; p = .000) of online learning platforms. Table 2.5 contains all the variances of mediating and endogenous variables of the TSM, including a summary of direct and indirect hypotheses and effect sizes and their critical ratios (CR).

A regional comparison between Eastern and Western Chinese university students was examined using two steps of invariance analyses (i.e., configural and metric invariance analyses) to validate our last hypothesis (*H7*). Before performing such analyses, this study cross-validated the TSM to observe the differences between Eastern (n_1 = 491) and Western (n_2 = 437) Chinese university students using two groups of samples. The results explained the validity of TSM for Eastern Chinese universities. For instance, the TSM for Eastern Chinese universities fit the data satisfactorily, with χ^2 = 426.073; *df* = 100; p = .000; RMSEA = .082; CFI = .956; and TLI = .947 (see Figure 2.4). The first six hypotheses (*H1–H6*) of the TSM were also valid for Eastern Chinese

Table 2.5 The Summary of TSM's Hypotheses

Hypotheses		β	CR (p)	Effect Size	Findings
Direct influences					
H1	CSE → PEU	.84	23.014 (0.000)	.844	Supported
H2	CSE → PU	.71	22.624 (0.000)	.713	Supported
H3	PEU → SAT	.27	10.177 (0.000)	.275	Supported
H4	PU → SAT	.71	23.936 (0.000)	.715	Supported
Indirect influences			χ^2 (p)		
H5	CSE → SAT (PU)	.504 (≥0.080)	8.425 (0.000)	.741	Supported
H6	CSE → SAT (PEU)	.226 (≥0.080)	15.442 (0.000)		Supported
Variances					
Mediating variables				PU	51%
				PEU	71%
Endogenous variables				SAT	82%

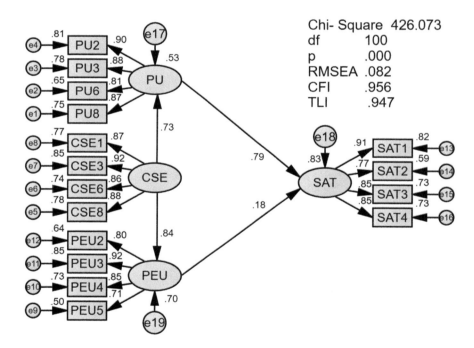

Figure 2.4 The TSM for Eastern Chinese universities.

universities, where CSE, PEU and PU explained 83% of the variability of students' satisfaction (SAT) with online learning platforms. Besides, CSE alone could explain 70% and 53% of the variance in PEU and PU of online learning platforms, respectively.

Testing Online Learning Satisfaction in Higher Education 55

Table 2.6 The Summary of TSM's Hypotheses for Eastern Chinese Universities

Hypotheses		β	CR (p)	Effect Size	Findings
Direct influences					
H1	CSE → PEU	.84	16.043 (0.000)	.837	Supported
H2	CSE → PU	.73	17.064 (0.000)	.728	Supported
H3	PEU → SAT	.18	5.077 (0.000)	.180	Supported
H4	PU → SAT	.79	18.869 (0.000)	.790	Supported
Indirect influences			χ^2 (p)		
H5	CSE → SAT (PU)	.576 (≥0.080)	11.715 (0.0)	.726	Supported
H6	CSE → SAT (PEU)	.151 (≥0.080)	4.236 (0.000)		Supported
Variances					
Mediating variables				PU	53%
				PEU	70%
Endogenous variables				SAT	83%

Table 2.6 contains all the variances of mediating and endogenous variables of the TSM for Eastern Chinese universities, including a summary of direct and indirect hypotheses and effect sizes and their critical ratios (CR).

In Table 2.7, this study reported 16 valid items and its factor loadings including Cronbach's alpha (α), mean (M) and standard deviation (SD) for Eastern Chinese universities.

The findings confirmed the validity of TSM for Western Chinese universities, which fit the data satisfactorily, with χ^2 = 404.744; df = 100; p = .000; RMSEA = .084; CFI = .951; and TLI = .941 (see Figure 2.5). The proposed six hypotheses (*H1–H6*) of the TSM were also found to be valid for Western Chinese universities, where exogenous (CSE) and mediating (PEU and PU) variables together explained 83% of the variance of students' satisfaction (SAT) with online learning platforms. CSE alone explained 72% and 48% of the variance in PEU and PU of online learning platforms, respectively. However, Figures 2.4 and 2.5 recognise several differences between the models for Eastern and Western Chinese universities in terms of their fit indices, loadings and error variances for items and variances, including path coefficients.

Table 2.8 contains all the variances of mediating and endogenous variables of the TSM for Western Chinese universities, including a summary of direct and indirect hypotheses and effect sizes and their critical ratios (CR).

In Table 2.9, this study reported 16 valid items and its factor loadings including Cronbach's alpha (α), mean (M) and standard deviation (SD) for Western Chinese universities.

In doing so, this study performed the invariance analyses after cross-validating the TSM to determine whether the aforementioned invariants of the causal structure of the TSM significantly moderate the relationships among the different facets. Our unconstrained models identified by TSM for Eastern Chinese universities and TSM for Western Chinese universities were grouped with the datasets (n_1 = 491 and n_2 = 437) to conduct configural analysis using

56 *Applying the Rasch Model and Structural Equation Modeling to Higher Education*

Table 2.7 Valid Parameters for Eastern Chinese Universities

Factors		Measurement Variables	Loadings	M	SD	α
CSE	CSE1	I am able to use the online learning platforms.	.87	4.963	.997	.936
	CSE3	I can navigate my way through the online learning platforms.	.92	4.774	1.078	
	CSE6	I have the ability to communicate with teachers and classmates through online learning platforms.	.86	4.741	1.082	
	CSE8	I can access the online learning platforms from home.	.88	4.927	1.023	
PEU	PEU2	I find it easy to access the online learning platforms.	.80	4.446	1.149	.889
	PEU3	It is easy for me to become skillful at using the online learning platforms.	.92	4.676	1.131	
	PEU4	It is easy for me to remember how to search and take courses by using the online learning platforms.	.85	4.601	1.179	
	PEU5	Interacting with the online learning platforms requires minimal mental effort.	.71	4.163	1.281	
PU	PU2	Using the online learning platforms helps me learn my courses.	.90	4.173	1.242	.922
	PU3	The online learning platforms address my study-related needs.	.88	4.303	1.170	
	PU6	Using the online learning platforms system allows me to accomplish more study work than would otherwise be possible.	.81	4.110	1.287	
	PU8	Using the online learning platforms enhances my knowledge and learning skills.	.87	4.191	1.179	
SAT	SAT1	Overall, I am satisfied with the ease of completing my tasks by using the online learning platforms.	.91	4.503	1.170	.911
	SAT2	The online learning platform service has greatly affected the way I learn.	.77	4.354	1.171	
	SAT3	The online learning platforms are indispensable and satisfactory services provided for Chinese university students.	.85	4.509	1.155	
	SAT4	Overall, I am satisfied with the amount of time it takes to complete my study tasks by using the online learning platforms.	.85	4.283	1.2823	

unstandardised estimates. The analyses showed that both models produced similar chi-square ($\chi^2 = 830.820$) and degree of freedom ($df = 200$) as required for configural analysis, and then we constrained all the paths of the model to estimate the metric invariance. The analyses showed that both models also

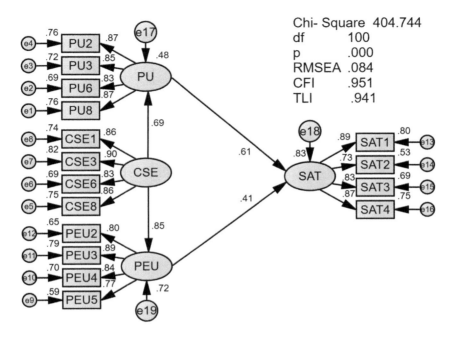

Figure 2.5 The TSM for Western Chinese universities.

Table 2.8 The Summary of TSM's Hypotheses for Western Chinese Universities

Hypotheses	β	CR (p)	Effect Size	Findings
Direct influences				
H1 CSE → PEU	.85	16.331 (0.000)	.846	Supported
H2 CSE → PU	.69	14.730 (0.000)	.692	Supported
H3 PEU → SAT	.41	10.002 (0.000)	.407	Supported
H4 PU → SAT	.61	14.755 (0.000)	.611	Supported
Indirect influences		χ^2 (p)		
H5 CSE → SAT (PU)	.420 (≥0.080)	10.009 (0.0)	.767	Supported
H6 CSE → SAT (PEU)	.348 (≥0.080)	8.236 (0.0)		Supported
Variances				
Mediating variables			PU	48%
			PEU	72%
Endogenous variables			SAT	83%

produced similar chi-square (χ^2 = 851.066) and degree of freedom (df = 204) as required for metric invariance analysis. Next, this study compared unconstrained and constrained models to compute the chi-squared differences and critical value (see Table 2.10). Based on these values, this research concluded that there is a significant regional difference between Eastern and Western

Table 2.9 Valid Parameters for Western Chinese Universities

Factors		Measurement Variables	Loadings	M	SD	α
CSE	CSE1	I am able to use the online learning platforms.	.86	4.730	.955	.924
	CSE3	I can navigate my way through the online learning platforms.	.90	4.483	1.063	
	CSE6	I have the ability to communicate with teachers and classmates through online learning platforms.	.83	4.579	1.045	
	CSE8	I can access the online learning platforms from home.	.86	4.494	1.035	
PEU	PEU2	I find it easy to access the online learning platforms.	.80	4.288	1.135	.895
	PEU3	It is easy for me to become skillful at using the online learning platforms.	.89	4.316	1.147	
	PEU4	It is easy for me to remember how to search and take courses by using the online learning platforms.	.84	4.378	1.109	
	PEU5	Interacting with the online learning platforms requires minimal mental effort.	.77	4.014	1.173	
PU	PU2	Using the online learning platforms helps me learn my courses.	.87	4.016	1.177	.915
	PU3	The online learning platforms address my study-related needs.	.85	4.066	1.124	
	PU6	Using the online learning platforms system allows me to accomplish more study work than would otherwise be possible.	.83	3.966	1.119	
	PU8	Using the online learning platforms enhances my knowledge and learning skills.	.87	3.984	1.109	
SAT	SAT1	Overall, I am satisfied with the ease of completing my tasks by using the online learning platforms.	.89	4.277	1.081	.903
	SAT2	The online learning platform service has greatly affected the way I learn.	.73	4.231	1.040	
	SAT3	The online learning platforms are indispensable and satisfactory services provided for Chinese university students.	.83	4.414	.990	
	SAT4	Overall, I am satisfied with the amount of time it takes to complete my study tasks by using the online learning platforms.	.87	4.297	1.046	

Testing Online Learning Satisfaction in Higher Education 59

Table 2.10 The Results of Regional Comparison

Models		Chi-Squared	df	Critical Value	Chi-Squared Change
Eastern and Western Chinese universities invariant of the TSM	Unconstrained	830.820	200	9.49 ($p > .05$)	20.246
	Constrained	851.066	204 4		

Chinese universities which moderates the relationships among the variables of TSM.

2.7 Discussion

In light of the TSM, this study tested and confirmed the first six hypotheses in the context of Chinese higher education and clarified the relationships among the exogenous (computer self-efficacy), endogenous (satisfaction) and mediating (perceived ease of use and usefulness) variables. Furthermore, this study revealed that there was a cross-regional invariant of the causal structure of the TSM between Eastern and Western Chinese university students. The findings obtained from the TSM have broadened the existing body of knowledge and current understanding of university students' satisfaction with using online learning platforms. Our findings also have implications for both theory and practice in terms of technology-enhanced online learning, especially during the COVID-19 pandemic.

Consistent with recent studies (Bin et al., 2020; Chen et al., 2019; Scherer et al., 2019; Thongsri et al., 2019; Yalçın & Kutlu, 2019), the statistical analyses have verified that Eastern and Western Chinese university students' computer self-efficacy directly impacts the perceived ease of use and usefulness of online learning platforms. This implies that the university students' perceived ease of use and usefulness of online learning platforms depend on their beliefs in their individual capabilities to use them for study. With the enhancement of computer self-efficacy, university students are likely to gradually accept the benefits and advantages of online learning platforms. Eventually, they will likely find the use of online learning platforms to be effortless. In fact, perceived ease of use and usefulness are frequently considered to be two crucial motivation variables in numerous TAM-based models. Considering the impact of computer self-efficacy on these variables, this study suggests it may be better to take such individual psychological factors into consideration when assessing new technology acceptance, adoption and satisfaction.

The direct influences of Eastern and Western Chinese university students' perceived ease of use and usefulness on their satisfaction with online learning platforms have been statistically confirmed. In fact, some up-to-date studies articulated that perceived ease of use and usefulness are associated with user satisfaction (Bin et al., 2020; Chen et al., 2019; Islam, 2016; Islam et al., 2018; Islam & Sheikh, 2020). This current study has validated such associations in measuring Chinese university students' satisfaction with online learning platforms. The findings of the TSM indicate that the easier university students find online learning platforms to use and the more benefits that online learning platforms provide, the more satisfied they will be. However, contradictory to Islam et al. (2018), this study revealed that perceived usefulness is relatively more effective than perceived ease of use in impacting university students' satisfaction.

The mediating roles played by perceived ease of use and usefulness between computer self-efficacy and satisfaction are also substantiated. These results are consistent with recent studies on measuring user satisfaction with the wireless internet (Islam et al., 2018), online research databases (Islam & Sheikh, 2020) and digital technologies (Bin et al., 2020) in higher and vocational and technical education. Specifically, although there are no direct relationships between computer self-efficacy and satisfaction, computer self-efficacy could improve university students' satisfaction by increasing their perception of ease of use and usefulness of online learning platforms. In this sense, computer self-efficacy is regarded as a distinct antecedent of the TSM. However, some researchers also mentioned that learners' intention to use may also multiply and mediate the relationships between satisfaction, perceived ease of use and usefulness (Bin et al., 2020; Chen et al., 2019). Further studies could also include intrinsic motivation variables, such as intention to use online learning platforms, when assessing university students' satisfaction.

As the invariance analyses exhibited, a cross-regional invariant of the causal structure of the TSM model between Eastern and Western Chinese university students do exist. This verifies that there is a significant difference between Eastern and Western Chinese university students' satisfaction with using online learning platforms. In other words, the direct influences of perceived ease of use and usefulness and the indirect influence of computer self-efficacy on university students' satisfaction with online learning platforms are moderated by regional factors. Previous studies claimed that the moderating effect of culture was significant on some paths of the TAM-based models (Jung & Lee, 2020) and that user satisfaction with ICTs differed in cross-cultural or cross-regional settings (Islam, 2016). This current study also implies that the generalizability of TSM may be constrained by region or culture. Just as Scherer and Teo (2019) pointed out, testing the structural invariance is critically important in interpreting possible cultural differences or similarities meaningfully. However, such cultural comparisons are rarely conducted, and such invariance is rarely examined (Scherer & Teo, 2019). Efforts have been made to narrow the gap in this current study, and this research suggests more data collection from universities in different cultures and regions. In addition, more invariance analyses in terms of cultural and regional differences should be encouraged in the future.

Based on our findings, three practical implications can be drawn. First of all, online learning platforms (e.g., LMSs) have been regarded as an integral part of the learning experience for students in many educational institutions in 2020–2021 (Turnbull et al., 2021). Considering that students have no other choice than online learning if they want to continue their studies during the COVID-19 pandemic, this study cannot overemphasise the importance of online learning platforms. Against this background, online learning platforms should undertake the significant responsibilities of serving and satisfying students, which are also recognised as the fundamental goals of promoting e-education in China (Ministry of Education, 2019). The TSM proves that perceived ease of use and usefulness are two important determinants of university students' satisfaction with online learning platforms. In view of this, on the one hand, online learning platforms need more simplified interfaces and registration and login systems to make them approachable to students. On the other hand, online learning platforms can still develop more useful features or learning support services to make them more beneficial to students. It is also necessary to publish relevant and better-designed guidebooks and manuals, with the help of which students will be able to use online learning platforms more easily and obtain more benefits from them. Course developers can also cooperate with platform designers to develop more accessible and beneficial programmes that target specific teaching contents. It is also worth mentioning that governments, universities and service providers can create social media for interactive communication with users which can help improve platform services. Moreover, the TSM indicates that computer self-efficacy is an influential factor of satisfaction that cannot be ignored. University students should gradually strengthen their basic computer competence in different ways so as to enhance their computer self-efficacy. Governments, universities and service providers can also hold lectures online to help university students improve computer capabilities. More importantly, students should also be encouraged to take the initiative in learning how to use platforms for deep online learning and learning management. Finally yet importantly, concerning regional differences, governments, universities and service providers can improve the quality of online learning platforms by taking into account the characteristics of student groups in different regions. Personalised online learning environments can be provided to satisfy the needs of different students and promote e-education equity.

2.8 Conclusion

There is an urgent need to measure learner satisfaction with using online learning platforms as millions of Chinese university students now rely on them to continue their studies due to the COVID-19 pandemic. In response to this need, the current study successfully applied the TSM and exhibited the direct and indirect impacts of computer self-efficacy, perceived ease of use and usefulness on university students' satisfaction with online learning platforms. It was found that Chinese university students are highly satisfied and that the TSM can powerfully explain and predict Chinese university students' satisfaction with online learning platforms. In particular, a regional comparison was conducted and the moderating effect of region on the paths of the TSM was examined statistically. The results indicated that there was a cross-regional invariant of the causal structure of the TSM between Eastern and Western Chinese university students. The current study can contribute to theoretical, methodical and practical understandings of university students' satisfaction with using online learning platforms, which have been recognised as irreplaceable emergency educational tools.

Despite the aforementioned discussion and conclusion, two limitations of this study should be acknowledged. On the one hand, this researchers selected five universities from two Eastern and Western Chinese provinces or municipalities. However, universities in other Eastern and Western Chinese provinces or municipalities were not included. Therefore, further studies can include more representative samples to validate our findings and increase the generalisability of the results. On the other hand, this was a quantitative study which employed a modeling test, whereas the qualitative method was ignored due to funding and time constraints. Based on our results, this research calls for more longitudinal studies which adopt a mixed method or triangulation method and are anchored in specific and detailed situations and contexts to explain university students' satisfaction with using online learning platforms.

2.9 Acknowledgements

This work was supported by the Peak Discipline Construction Project of Education at East China Normal University and Fundamental Research Funds for the Central Universities (2020ECNU-HLYT035).

2.10 References

Abdullah, F., & Ward, R. (2016). Developing a general extended technology acceptance model for e-learning (GETAMEL) by analysing commonly used external factors. *Computers in Human Behavior, 56*, 238–256.

Ajzen, I., & Fishbein, M. (1980). *Understanding attitudes and predicting social behavior.* Prentice Hall.

Al-Fraihat, D., Joy, M., Masa'Deh, R., & Sinclair, J. (2020). Evaluating e-learning systems success: An empirical study. *Computers in Human Behavior, 102*, 67–86.

Ameen, N., Willis, R., Abdullah, M. N., & Shah, M. (2019). Towards the successful integration of e-learning systems in higher education in Iraq: A student perspective. *British Journal of Educational Technology, 50*(3), 1434–1446.

Bandura, A. (1977). Self-efficacy: Toward a unifying theory of behavioral change. *Psychological Review, 84*(2), 191–215.

Bandura, A. (1997). *Self-efficacy: The exercise of control.* Freeman.

Barclay, C., Donalds, C., & Osei-Bryson, K. (2018). Investigating critical success factors in online learning environments in higher education systems in the Caribbean. *Information Technology for Development, 24*(3), 582–611.

Bin, E., Islam, A. Y. M. A., Gu, X., Spector, J. M., & Wang, F. (2020). A study of Chinese technical and vocational college teachers' adoption and gratification in new technologies. *British Journal of Educational Technology.* https://doi.org/10.1111/bjet.12915

Bond, T. G., & Fox, C. M. (2001). *Applying the Rasch model: Fundamental measurement in the human science*. Lawrence Erlbaum.

Burkell, J. (2003). The dilemma of survey nonresponse. *Library and Information Science Research*, *25*(3), 239–263.

Cai, Q., Yuan, Z., & He, W. (2021). The realistic dilemma, logical essentials and solutions of the overall revitalization of higher education in western China. *University Education Science*, *28*(1), 26–35.

Chen, H., Islam, A. Y. M. A., Gu, X., Teo, T., & Peng, Z. (2019). Technology-enhanced learning and research using databases in higher education: The application of the ODAS model. *Educational Psychology*. http://doi.org/10.1080/01443410.2019.1614149

Chingos, M. M., Griffiths, R. J., Mulhern, C., & Spies, R. R. (2017). Interactive online learning on campus: Comparing students' outcomes in hybrid and traditional courses in the university system of Maryland. *The Journal of Higher Education*, *88*(2), 210–233.

Cidral, W. A., Oliveira, T., Di Felice, M., & Aparicio, M. (2018). E-learning success determinants: Brazilian empirical study. *Computers & Education*, *122*, 273–290.

Coates, H., James, R., & Baldwin, G. (2005). A critical examination of the effects of learning management systems on university teaching and learning. *Tertiary Education and Management*, *11*(1), 19–36. https://doi.org/10.1007/s11233-004-3567-9

Dai, H. M., Teo, T., Rappa, N. A., & Huang, F. (2020). Explaining Chinese university students' continuance learning intention in the MOOC setting: A modified expectation confirmation model perspective. *Computers & Education*, *150*, 1–16.

Davis, F. D., Bagozzi, R. P., & Warshaw, P. R. (1989). User acceptance of computer-technology: A comparison of two theoretical models. *Management Science*, *35*(8), 982–1003.

de Souza Rodrigues, M. A., Chimenti, P., & Nogueira, A. R. R. (2021). An exploration of eLearning adoption in the educational ecosystem. *Education and Information Technologies*, *26*(1), 585–615. https://doi.org/10.1007/s10639-020-10276-3

Deng, R., Benckendorff, P., & Gannaway, D. (2019). Progress and new directions for teaching and learning in MOOCs. *Computers & Education*, *129*, 48–60.

Dong, Y., Xu, C., Chai, C. S., & Zhai, X. (2020). Exploring the structural relationship among teachers' technostress, technological pedagogical content knowledge (TPACK), computer self-efficacy and school support. *The Asia-Pacific Education Researcher*, *29*(2), 147–157.

Esteban-Millat, I., Martinez-Lopez, F. J., Pujol-Jover, M., Gazquez-Abad, J. C., & Alegret, A. (2018). An extension of the technology acceptance model for online learning environments. *Interactive Learning Environment*, *26*(7), 895–910.

Etikan, I., Musa, S. A., & Alkassim, R. S. (2016). Comparison of convenience sampling and purposive sampling. *American Journal of Theoretical and Applied Statistics*, *5*(1), 1–4. https://doi.org/10.11648/j.ajtas.20160501.11

Farhan, W., Razmak, J., Demers, S., & Laflamme, S. (2019). E-learning systems versus instructional communication tools: Developing and testing a new e-learning user interface from the perspectives of teachers and students. *Technology in Society*, *59*, 1–12.

Fırat, M., Kılınç, H., & Yüzer, T. V. (2018). Level of intrinsic motivation of distance education students in e-learning environments. *Journal of Computer Assisted Learning*, *34*(1), 63–70.

Fornell, C., & Larcker, D. F. (1981). Evaluating structural equation models with unobservable variables and measurement error. *Journal of Marketing Research*, *48*, 39–50.

Granić, A., & Marangunić, N. (2019). Technology acceptance model in educational context: A systematic literature review. *British Journal of Educational Technology*, *50*(5), 2572–2593.

Habib, M. N., Jamal, W., Khalil, U., & Khan, Z. (2021). Transforming universities in interactive digital platform: Case of city university of science and information technology. *Education and Information Technologies*, *26*(1), 517–541. https://doi.org/10.1007/s10639-020-10237-w

Hair, J. F., Black, W., Babin, B. J., & Anderson, R. E. (2010). *Multivariate data analysis: A global perspective.* Pearson.

Hassan, H. E., & Wood, V. R. (2020). Does country culture influence consumers' perceptions toward mobile banking? A comparison between Egypt and the United States. *Telematics and Informatics, 46,* 1–14.

Heckel, C., & Ringeisen, T. (2019). Pride and anxiety in online learning environments: Achievement emotions as mediators between learners' characteristics and learning outcomes. *Journal of Computer Assisted Learning, 35*(5), 667–677.

Herrador-Alcaide, T. C., Hernández-Solís, M., & Galván, R. S. (2019). Feelings of satisfaction in mature students of financial accounting in a virtual learning environment: An experience of measurement in higher education. *International Journal of Educational Technology in Higher Education, 16,* 1–19.

Hu, L. T., & Bentler, P. M. (1999). Cutoff criteria for fit indexes in covariance structure analysis: Conventional criteria versus new alternatives. *Structural Equation Modeling, 6*(1), 1–55.

Hu, X., & Xie, Z. (2020). On the advantages and challenges of online teaching & learning in universities & colleges under the epidemic. *China Higher Education Research, 36*(4), 18–22.

Huang, L. K. (2017). A cultural model of online banking adoption: Long-term orientation perspective. *Journal of Organizational and End User Computing, 29*(1), 1–22.

Ilgaz, H., & Gülbahar, Y. (2015). A snapshot of online learners: E-readiness, e-satisfaction and expectations. *International Review of Research in Open & Distributed Learning, 16*(2), 171–187.

Isaac, O., Aldholay, A., Abdullah, Z., & Ramayah, T. (2019). Online learning usage within Yemeni higher education: The role of compatibility and task-technology fit as mediating variables in the IS success model. *Computers & Education, 136,* 113–129.

Islam, A. Y. M. A. (2011). *Online database adoption and satisfaction model.* Lambert Academic Publishing.

Islam, A. Y. M. A. (2014). Validation of the technology satisfaction model (TSM) developed in higher education: The application of structural equation modeling. *International Journal of Technology and Human Interaction, 10*(3), 44–57.

Islam, A. Y. M. A. (2016). Development and validation of the technology adoption and gratification (TAG) model in higher education: A cross-cultural study between Malaysia and China. *International Journal of Technology and Human Interaction, 12*(3), 78–105.

Islam, A. Y. M. A., Leng, C. H., & Singh, D. (2015). Efficacy of the technology satisfaction model (TSM): An empirical study. *International Journal of Technology and Human Interaction, 11*(2), 45–60.

Islam, A. Y. M. A., Mok, M. M. C., Qian, X., & Leng, C. H. (2018). Factors influencing students' satisfaction in using wireless internet in higher education: Cross-validation of TSM. *The Electronic Library, 36*(1), 2–20.

Islam, A. Y. M. A., & Sheikh, A. (2020). A study of the determinants of postgraduate students' satisfaction of using online research databases. *Journal of Information Science, 46*(2), 273–287.

Jiang, Z., & Zhao, C. (2018). Learner satisfaction: The ultimate destination of teacher support behavior in online learning. *Modern Distance Education, 35*(6), 51–59.

Jung, I., & Lee, J. (2020). A cross-cultural approach to the adoption of open educational resources in higher education. *British Journal of Educational Technology, 51*(1), 263–280.

Kingsbury, I. (2021). Online learning: How do brick and mortar schools stack up to virtual schools? *Education and Information Technologies.* https://doi.org/10.1007/s10639-021-10450-1

Lane, A., Caird, S., & Weller, M. (2015). The potential social, economic and environmental benefits of MOOCs: Operational and historical comparisons with a massive "Closed Online" course. *Distance Education in China, 35*(2), 18–24.

Liu, X., & Boone, W. (2006). *Application of Rasch measurement in science education.* JAM Press.

Machado-Da-Silva, F. N., Meirelles, F. D. S., Filenga, D., & Filho, M. B. (2014). Student satisfaction process in virtual learning system: Considerations based in information and service quality from Brazil's experience. *Turkish Online Journal of Distance Education, 15*(3), 122–142.

Martins, J., Branco, F., Gonçalves, R., Au-Yong-Oliveira, M., Oliveira, T., Naranjo-Zolotov, M., & Cruz-Jesus, F. (2018). Assessing the success behind the use of education management information systems in higher education. *Telematics and Informatics, 38*(1), 182–193.

Ministry of Education of the People's Republic of China. (2019). *Guidelines on promoting the healthy development of e-education of the ministry of education and ten other government departments.* Retrieved April 29, 2020, from www.moe.gov.cn/srcsite/A03/moe_1892/moe_630/201909/t20190930_401825.html

Ministry of Education of the People's Republic of China. (2020a). *Making use of online platforms: "Suspending Classes without Stopping Learning".* Retrieved April 26, 2020, from www.moe.gov.cn/jyb_xwfb/gzdt_gzdt/s5987/202001/t20200129_416993.html

Ministry of Education of the People's Republic of China. (2020b). *Online teaching practice has achieved good results during the pandemic in universities across the country, and the Ministry of Education will launch international platforms for online teaching in universities.* Retrieved April 26, 2020, from www.moe.gov.cn/jyb_xwfb/gzdt_gzdt/s5987/202004/t20200410_442294.html

Ministry of Education of the People's Republic of China. (2020c). *Notice of the higher education department of the Ministry of Education on continuing to organize online platforms to provide resources and service programs to support e-teaching in universities during the epidemic prevention and control period.* Retrieved April 26, 2020, from www.moe.gov.cn/s78/A08/A08_gggs/s8468/202002/t20200206_418504.html

Ministry of Education of the People's Republic of China. (2020d). *During the epidemic prevention and control, the online teaching organization and management of colleges and universities should be done well.* Retrieved April 26, 2020, from www.moe.gov.cn/jyb_xwfb/gzdt_gzdt/s5987/202002/t20200205_418131.html

Mpungose, C. B. (2020). Is Moodle or WhatsApp the preferred e-learning platform at a South African university? First-year students' experiences. *Education and Information Technologies, 25*(2), 927–941. https://doi.org/10.1007/s10639-019-10005-5

National Bureau of Statistics of China. (2019). *China statistical yearbook.* Retrieved March 31, 2020, from www.stats.gov.cn/tjsj/ndsj/2019/indexch.htm

Pham, L., Limbu, Y. B., Bui, T. K., Nguyen, H. T., & Pham, H. T. (2019). Does e-learning service quality influence e-learning student satisfaction and loyalty? Evidence from Vietnam. *International Journal of Educational Technology in Higher Education, 16*, 1–26.

Pinho, C., Franco, M., & Mendes, L. (2021). Application of innovation diffusion theory to the E-learning process: Higher education context. *Education and Information Technologies, 26*(1), 421–440. https://doi.org/10.1007/s10639-020-10269-2

Rasch, G. (1960). *Probabilistic models for some intelligence and attainment tests.* Nielsen & Lydiche.

Salam, M., & Farooq, M. S. (2020). Does sociability quality of web-based collaborative learning information system influence students' satisfaction and system usage? *International Journal of Educational Technology in Higher Education, 17*, 1–39.

Scherer, R., Siddiq, F., & Tondeur, J. (2019). The technology acceptance model (TAM): A meta-analytic structural equation modeling approach to explaining teachers' adoption of digital technology in education. *Computers & Education, 128*, 13–35.

Scherer, R., & Teo, T. (2019). Editorial to the special section – technology acceptance models: What we know and what we (still) do not know. *British Journal of Educational Technology, 50*(5), 2387–2393.

Sobel, M. E. (1982). Asymptotic confidence intervals for indirect effects in structural equation models. *Sociological Methodology, 13*, 290–312.

Su, C., & Chen, C. (2020). Investigating university students' attitude and intention to use a learning management system from a self-determination perspective. *Innovations in Education and Teaching International.* https://doi.org/10.1080/14703297.2020.1835688

Thongsri, N., Shen, L., & Bao, Y. (2019). Investigating academic major differences in perception of computer self-efficacy and intention toward e-learning adoption in China. *Innovations in Education and Teaching International.* https://doi.org/10.1080/14703297.2019.1585904

Turnbull, D., Chugh, R., & Luck, J. (2019). Learning management systems: An overview. In A. Tatnall (Ed.), *Encyclopedia of education and information technologies.* Springer. https://doi.org/10.1007/978-3-030-10576-1_248

Turnbull, D., Chugh, R., & Luck, J. (2020). Learning management systems: A review of the research methodology literature in Australia and China. *International Journal of Research & Method in Education.* https://doi.org/10.1080/1743727X.2020.1737002.

Turnbull, D., Chugh, R., & Luck, J. (2021). Issues in learning management systems implementation: A comparison of research perspectives between Australia and China. *Education and Information Technologies.* https://doi.org/10.1007/s10639-021-10431-4

UNESCO. (2020a). *COVID-19 educational disruption and response.* Retrieved May 4, 2020, from https://en.unesco.org/covid19/educationresponse

UNESCO. (2020b). *Half of world's student population not attending school: UNESCO launches global coalition to accelerate deployment of remote learning solutions.* Retrieved May 12, 2020, from https://en.unesco.org/news/half-worlds-student-population-not-attending-school-unesco-launches-global-coalition-accelerate

Uppal, M. A., Ali, S., & Gulliver, S. R. (2018). Factors determining e-learning service quality. *British Journal of Educational Technology, 49*(3), 412–426. https://doi.org/10.1111/bjet.12552

Vasileva-Stojanovska, T., Malinovski, T., Vasileva, M., Jovevski, D., & Trajkovik, V. (2015). Impact of satisfaction, personality and learning style on educational outcomes in a blended learning environment. *Learning and Individual Differences, 38*, 127–135.

Venkatesh, V., & Davis, F. D. (1996). A model of the antecedents of perceived ease of use: Development and test. *Decision Sciences, 27*(3), 451–481.

Virtanen, M. A., Kääriäinen, M., Liikanen, E., & Haavisto, E. (2017). The comparison of students' satisfaction between ubiquitous and web-based learning environments. *Education and Information Technologies, 22*(5), 2565–2581. https://doi.org/10.1007/s10639-016-9561-2

Wang, H., Lin, V., Hwang, G., & Liu, G. (2019). Context-aware language-learning application in the green technology building: Which group can benefit the most? *Journal of Computer Assisted Learning, 35*(3), 359–377.

Wang, J., Cui, Y., & Yan, Y. (2020). Analysis on effect and potential influence factors of massive online education practice during the COVID-19. *E-Education Research, 41*(6), 5–12.

Xu, Y., & Yao, J. (2018). Could online education promote education equity? An empirical study based on online public elective courses. *E-Education Research, 39*(4), 38–45.

Xu, Y., & Ye, X. (2018). MOOCs promote equity in education: Reality or illusion? *Modern Distance Education Research, 25*(3), 83–93.

Yalçın, M. E., & Kutlu, B. (2019). Examination of students' acceptance of and intention to use learning management systems using extended TAM. *British Journal of Educational Technology*, *50*(5), 2414–2431.

Yang, J. (2020). Grasping the opportunities of resource opening and promoting the learning from students to teachers: The practice and thinking of "Suspending Classes without Stopping Learning" in underdeveloped frontier mountainous areas and counties. *China Educational Technology*, *27*(4), 29–31.

Yen, S., Lo, Y., Lee, A., & Enriquez, J. (2018). Learning online, offline, and in-between: Comparing student academic outcomes and course satisfaction in face-to-face, online, and blended teaching modalities. *Education and Information Technologies, 23*(5), 2141–2153. https://doi.org/10.1007/s10639-018-9707-5

Yuen, A. H. K., Cheng, M., & Chan, F. H. F. (2019). Student satisfaction with learning management systems: A growth model of belief and use. *British Journal of Educational Technology*, *50*(5), 2520–2535.

Yunusa, A. A., & Umar, I. N. (2021). A scoping review of Critical Predictive Factors (CPFs) of satisfaction and perceived learning outcomes in E-learning environments. *Education and Information Technologies, 26*(1), 1223–1270. https://doi.org/10.1007/s10639-020-10286-1

Zhang, M. H., Su, C. Y., Li, Y., & Li, Y. Y. (2020). Factors affecting Chinese university students' intention to continue using virtual and remote labs. *Australasian Journal of Educational Technology*, *36*(2), 169–185. https://doi.org/10.14742/ajet.5939

Zhu, J., & Mok, M. M. C. (2020). Predictors of students' participation in internet or computer tutoring for additional instruction and its effect on academic achievement. *Journal of Computer Assisted Learning*. https://doi.org/10.1111/jcal.12440

Zhu, Z., Guo, S., Wu, D., & Liu, S. (2020). Policy interpretation, key issues and countermeasures of "Suspending Classes without Stopping Learning". *China Educational Technology*, *27*(4), 1–7.

Zhu, Z., & Peng, H. (2020). Omnimedia learning ecology: A practical solution to cope with schooling difficulties during a large-scale epidemic. *China Educational Technology*, *27*(3), 1–6.

Appendix

The Survey Questionnaire (English and Chinese versions)

This questionnaire attempts to investigate "**Testing Online Learning Satisfaction in Higher Education**"
评价高等教育的网络学习满意度

Direction

说明

The questionnaire has five sections. For all multiple-choice questions, please indicate your response by placing a tick (/) in the appropriate box. For six-point Likert scale questions, please circle the number of your choice. If you wish to comment on any question or qualify your answer, please feel free to use the space in the margin or write your comments on a separate sheet of paper. All information that you provide will be kept strictly confidential. Thank you!

您好！

欢迎参加"高校运用网络平台提升学习"的调查工作。此次调查是本学习团队为了提升网络学习平台的建设而设计，旨在通过关于我校学生对网络学习平台的使用现状的调查为网络学习平台建设提出改进意见。希望您能抽出一点时间积极配合我们的调查工作，谢谢您的参与。

Section I: Demographic information

第一部分：基本信息

1. Gender: Male ☐
           Female ☐

1. 性别: 男 ☐
        女 ☐

2. Age: 17–18 years old   ☐
            19–20 years old   ☐
            21–22 years old   ☐
            23–24 years old   ☐
            25 years old and above ☐

2. 年龄： 17–18 岁   ☐
            19–20 岁   ☐
            21–22 岁   ☐
            23–24 岁   ☐
            25岁及以上   ☐

3. Number of years studying at the college/university
    a.   0–1 years
    b.   2–3 years
    c.   4–5 years
    d.   6–7 years
    e.   8 years and above

3. 在大学（包括本科）就读的时间：
    a.   0–1 年
    b.   2–3 年
    c.   4–5 年
    d.   6–7 年
    e.   8 年及以上

4. Place of your college_____
4. 您就读学校所在省份：_____
5. Place of birth _____
5. 出生地省份：_____
6. Academic achievement score (CGPA/GPA) _____
6. 绩点：_____

Section II: Perceived ease of use of the online learning platforms

第二部分 对网络学习平台的感知易用性

问题1–10，请在相应的**选项**上用(√)标出您的选择：

For questions 1–10, rate how much you agree with each statement using the following scale:

1 = Strongly Disagree (SD)   4 = Slightly Agree (SLA)
1 = 非常不同意   4 = 有些同意
2 = Moderately Disagree (MD)   5 = Moderately Agree (MA)
2 = 比较不同意   5 = 比较同意
3 = Slightly Disagree (SLD)   6 = Strongly Agree (SA)
3 = 有些不同意   6 = 非常同意

No. 序号	Items 选项	SD 非常不同意	MD 比较不同意	SLD 有些同意	SLA 有些同意	MA 比较同意	SA 非常同意
1	I find the online learning platforms easy to use. 网络学习平台使用起来很方便	1	2	3	4	5	6
2	I find it easy to access the online learning platforms. 访问网络学习平台很便捷	1	2	3	4	5	6
3	It is easy for me to become skillful at using the online learning platforms. 我可以轻松、熟练地使用网络学习平台	1	2	3	4	5	6
4	It is easy for me to remember how to search and take courses by using the online learning platforms. 我能轻松回想起如何使用网络学习平台搜索和学习课程	1	2	3	4	5	6
5	Interacting with the online learning platforms requires minimal mental effort. 在与网络学习平台界面进行互动时，我感到毫不费劲	1	2	3	4	5	6
6	I do not feel the need to consult any user manual when using the online learning platforms. 在使用网络学习平台时，我不需要求助在线助手或使用说明	1	2	3	4	5	6
7	I find it easy to get the online learning platforms to help me take the courses. 我觉得使用网络学习平台帮助我开展课程学习是十分轻松	1	2	3	4	5	6
8	My interaction with the online learning platforms is within my comprehension. 与网络学习平台界面的互动方式和要求是在我理解范围内的	1	2	3	4	5	6
9	Interacting with the online learning platforms is very stimulating for me. 与网络学习平台界面互动，对我具有促进作用	1	2	3	4	5	6
10	I find it easy to choose courses of different categories (for example, education, engineering and business.). 我觉得很容易就可以通过网络学习平台的分类找到相应的课程（比如：教育类/工程类/商务类/语言类等）	1	2	3	4	5	6

Section III: Perceived usefulness of the online learning platforms

第三部分 对网络学习平台的感知有用性

问题1–10，请在相应的选项用(√)标出您的选择：

For questions 1–10, rate how much you agree with each statement using the following scale:

1 = Strongly Disagree (SD)　　　4 = Slightly Agree (SLA)
1 = 非常不同意　4 = 有些同意
2 = Moderately Disagree (MD)　　5 = Moderately Agree (MA)
2 = 比较不同意　5 = 比较同意
3 = Slightly Disagree (SLD)　　　6 = Strongly Agree (SA)
3 = 有些不同意　　　　　　　　6 = 非常同意

No. 序号	Items 选项	SD 非常不同意	MD 比较不同意	SLD 有些同意	SLA 有些同意	MA 比较同意	SA 非常同意
1	Using the online learning platforms enables me to get the courses I am interested in. 使用网络学习平台有助于我学习感兴趣的课程	1	2	3	4	5	6
2	Using the online learning platforms helps me learn my courses. 使用网络学习平台有助于我更好地学习我的课程	1	2	3	4	5	6
3	The online learning platforms address my study-related needs. 网络学习平台能够满足我学习相关的需求	1	2	3	4	5	6
4	My study would be difficult to perform without the online learning platforms. 如果没有网络学习平台，我的学习会很难进行	1	2	3	4	5	6
5	Using the online learning platforms system saves me time. 使用网络学习平台节省了我的时间	1	2	3	4	5	6
6	Using the online learning platforms system allows me to accomplish more study work than would otherwise be possible. 与采用其他方式相比，使用网络学习平台帮助我完成更多的学习工作	1	2	3	4	5	6

(Continued)

72 *Applying the Rasch Model and Structural Equation Modeling to Higher Education*

(Continued)

No. 序号	Items 选项	SD 非常不 同意	MD 比较 不同意	SLD 有些 同意	SLA 有些 同意	MA 比较 同意	SA 非常 同意
7	Using the online learning platforms improves the quality of my study. 使用网络学习平台提高了我的学习质量	1	2	3	4	5	6
8	Using the online learning platforms enhances my knowledge and learning skills. 使用网络学习平台提升了我的知识与学习能力	1	2	3	4	5	6
9	Using the online learning platforms improves my study performance. 使用网络学习平台有助于我提高学习成绩	1	2	3	4	5	6
10	Using the online learning platforms provides me with the latest information on particular areas of study. 网络学习平台为我提供了特定学习领域的最新信息	1	2	3	4	5	6

Section IV: Computer self-efficacy of the online learning platforms

第四部分：自我效能

问题1–8，请在相应的**选项**上用(√)标出您的选择：

For questions 1–8, rate how much you agree with each statement using the following scale:

1 = Strongly Disagree (SD)　　4 = Slightly Agree (SLA)
1 = 非常不同意　　　　　　　4 = 有些同意
2 = Moderately Disagree (MD)　5 = Moderately Agree (MA)
2 = 比较不同意　　　　　　　5 = 比较同意
3 = Slightly Disagree (SLD)　　6 = Strongly Agree (SA)
3 = 有些不同意　　　　　　　6 = 非常同意

No. 序号	Items 选项	SD 非常不 同意	MD 比较 不同意	SLD 有些 同意	SLA 有些 同意	MA 比较 同意	SA 非常 同意
1	I am able to use the online learning platforms. 我知道如何使用网络学习平台	1	2	3	4	5	6

No. 序号	Items 选项	SD 非常不 同意	MD 比较 不同意	SLD 有些 同意	SLA 有些 同意	MA 比较 同意	SA 非常 同意
2	I can download learning materials from the online learning platforms. 我知道如何下载网络学习平台中的学习材料	1	2	3	4	5	6
3	I can navigate my way through the online learning platforms. 我可以自如地使用网络学习平台	1	2	3	4	5	6
4	I can use the online learning platforms to enhance the quality of my study. 我能够使用网络学习平台来提升我的学习质量	1	2	3	4	5	6
5	I know how to bookmark courses in the online learning platform so that I can easily find them out next time. 我知道如何在在线学习平台上为课程添加书签，这样下次我就可以轻松地找到他们	1	2	3	4	5	6
6	I have the ability to communicate with teachers and classmates through online learning platforms. 我知道如何通过网络学习平台与老师和同学交流	1	2	3	4	5	6
7	I have the knowledge and skills required to benefit from using the online learning platforms. 我拥有有效利用网络学习平台的知识和技能	1	2	3	4	5	6
8	I can access the online learning platforms from home. 我知道如何从家里访问网络学习平台	1	2	3	4	5	6

Section V: Satisfaction of the online learning platforms

第五部分： 对网络学习平台的满意度

问题1-5，请在相应的**选项**上用(√)标出您的选择：

For questions 1–5, rate the likelihood of each using the following scale:

1 = Very Dissatisfied (VD)   4 = Slightly Satisfied (SS)
1 = 非常不满意   4 = 有些满意
2 = Moderately Dissatisfied (MD)   5 = Moderately Satisfied (MS)
2 = 比较不满意   5 = 比较满意
3 = Slightly Dissatisfied (SD)   6 = Very Satisfied (VS)
3 = 有些不满意   6 = 非常满意

No. 序号	Items 选项	VD 非常不满意	MD 比较不满意	SD 有些不满意	SS 有些满意	MS 比较满意	VS 非常满意
1	Overall, I am satisfied with the ease of completing my task by using the online learning platforms. 总体而言，我对使用网络学习平台来完成相应任务的便利性表示满意	1	2	3	4	5	6
2	The online learning platforms service has greatly affected the way I learn. 学校的网络学习平台已经很大程度上影响了我学习的方式	1	2	3	4	5	6
3	The online learning platforms are indispensable and satisfactory services provided for Chinese college students. 网络学习平台资源对大学生来说是不可或缺的、令人满意的服务资源	1	2	3	4	5	6
4	Overall, I am satisfied with the amount of time it takes to complete my study task by using the online learning platforms. 总体而言，我对利用网络学习平台来完成学习任务所花费的时间量表示满意	1	2	3	4	5	6
5	I am satisfied with the structure of accessible information (available as categories of courses, or by date of courses) of the online learning platforms. 对网络学习平台提供的可访问资源的分类布局（按课程类别、按课程更新时间），我表示满意	1	2	3	4	5	6

3 Assessing Online Research Databases in Higher Education Using the TSM

3.1 Introduction

The invention of the World Wide Web had a remarkable effect on academic libraries. Today, libraries are no longer just repositories of books, research collections and other printed materials. They are now electronic gateways to ostensibly unlimited amounts of information (Byerley et al., 2007; Katabalwa, 2016). In addition, the internet offers exciting opportunities to access information (Li et al., 2022; Islam et al., 2018; Dukić, 2014; Horwath, 2002). Similarly, advancements in information and communication technologies (ICTs) have altered the way information is generated, stored and accessed (Islam et al., 2019), while the internet and web-based technologies have facilitated the publication and dissemination of information in digital formats (Ahmed, 2013; Ayoo & Lubega, 2014; Dermody & Majekodunmi, 2011). These technological advancements have resulted in the growth of an entire new array of information services (Khan et al., 2009). Both academic (Du et al., 2022) and public libraries have responded to these trends by strengthening their existing services and expanding their collections of online resources (Aagaard & Arguello, 2015). The academic library is a vital platform for obtaining a broad array of resources and the latest information to support teaching (Mehta & Wang, 2020), doing research (Chen et al., 2020; Islam et al., 2021) and university development. Online databases are one of the major achievements of this technological revolution. Within a short period of time, they have become a key source of current and reliable information for a range of users in all scientific areas (Dukić, 2013).

According to Ani and Ahiauzu (2008), "online databases are collections of electronic information sources from various fields and disciplines". For Koprivica and Grabovac (2010), "online databases are organised sets of data, in which each unit is marked and described in the same way". Online databases can also be defined as "electronic collections of information accessible via the Internet, often containing journal articles or references to such articles" (Guruprasad et al., 2012). In addition, "online databases are organised sets of scholarly and professional publications in an electronic form accessed through a computer network". Islam (2011) defines online databases as the indexes of

DOI: 10.1201/9781003384724-3

electronic journals that facilitate the tracing and retrieving of journal articles via the internet. These databases may include full-text articles or only citation abstracts (Tripathi & Kumar, 2014). Academic libraries are investing heavily in subscribing to full-text research databases to supplement their existing collections (Collins, 2012; Chen et al., 2020; Islam, 2011; Islam et al., 2015).

The COMSATS University Islamabad (CUI) library recognises the value of online research databases. Consequently, the library provides access to 28 research journal databases. CUI accesses these databases through the Higher Education Commission National Digital Library (HEC NDL). The HEC in Pakistan is responsible for controlling the country's higher education system and plays a vital role in making these databases available to all public and private universities. The HEC NDL is an HEC programme to build a knowledge-based economy in Pakistan (Said, 2006) by providing all researchers within the country's higher education system with access to internationally published peer-reviewed journal literature in all disciplines. The HEC started to offer these databases in 2004 to boost the research productivity of Pakistan's universities (Pakistan HEC National Digital Library, 2017).

CUI users can access these databases at their campuses by using their laptops and mobile phones and through computer workstations installed in the library. Moreover, off-campus access is also offered to the library users through VPN accounts. Certainly, these databases are contributing considerably towards quality research work at CUI. Today, CUI is one of the top research-oriented institutions in Pakistan. CUI's online databases have been accessible to users for more than a decade, but their usage and satisfaction among the university population, especially among postgraduate students, is unknown.

Hence, there is a need to understand the determinants of postgraduate students' satisfaction with online research databases using the technology satisfaction model in an academic library. Moreover, validating this model will provide academic librarians with a tool to assess users' satisfaction with technology. This paper offers a review of the literature pertinent to online research databases. It subsequently discusses the methodology and findings of this study. It concludes with the limitations of this study and offers recommendations for future research.

3.2 Review of the Literature

Academic librarians frequently guide library users in their quest for relevant information materials. As the quantity of available information continues to grow, the task of retrieving relevant information becomes more and more challenging (Bates et al., 2017). Academic libraries all over the world are experiencing a paradigm shift in acquiring, storing and retrieving information, because of the application of information communication technologies. Libraries organised with ICTs can offer friendlier information retrieval services to users (Tlakula & Fombad, 2017; Tripathi & Kumar, 2014; Ukachi, 2015). In the current technology-rich educational environment, most of the information is stored and retrieved with the help of online databases. Consequently, online databases have become the fundamental element of library services (Stewart et al., 2005). Online databases are extremely useful for searching huge volumes of information within a minimum amount of time (Singh & Gautam, 2004). The majority of universities and institutions provide online databases to their scholars to aid academic and research work (Khan et al., 2009; Sinh & Nhung, 2012; Du et al., 2022; Chen et al., 2020; Islam et al., 2015; Islam, 2011). Online databases in libraries first appeared as single-user CD-ROM workstations in the 1980s but became commonplace with the explosive growth of internet technology in the 1990s and 2000s. Easy remote access, advanced searching functionality and formerly unavailable premium content make these resources attractive to users in both academic and public libraries (Albitz, 2008; Atakan et al., 2008; Byerley et al., 2007).

In this age of information, the widespread use of the internet and communication technologies has transformed how information is produced and disseminated (Russell, 2008). The volume and complexity of the available information also require users to have some basic knowledge and skills to be able to effectively locate and retrieve information (Brar, 2015). In other words, information users need to be information literate (Sinha, 2016). A person can be called information literate when he or she is able to find, use and synthesise information by himself or herself. It means that a digitally literate person can use, manage, quote and share sources of digital information in a more efficient manner (Jeffrey et al., 2011). The knowledge and skill of information users, the ability to select appropriate search strategies and the effective use of information sources is called information literacy. Digital information literacy is a component of information literacy (Brar, 2015). In simple words, digital information literacy is the ability of an individual to properly use and evaluate digital resources, tools and services (Brar, 2015). Thus, digital information literacy and the skills and competencies of library users have great implications for the use of digital resources (Li, 2007). Many previous studies (Islam, 2011; Islam et al., 2015) have also illustrated the significant role of and relationship between computer self-efficacy and user satisfaction in using online research databases.

At present, universities throughout the world are allocating substantial budgets to subscribe to costly online research databases (Kandasamy & Vinitha, 2014; Verma, 2016; Islam, 2011; Islam et al., 2015; Chen et al., 2020; Du et al., 2022). The Junaid Zaidi Library at CUI, for example, spends 50% of its annual budget on its subscriptions to research databases. Therefore, exploring university students' usage and their satisfaction with these databases can benefit libraries. As user satisfaction is the ultimate goal of libraries, accordingly many researchers, including Peritz et al. (2003), Coombs (2005), Falk (2005), Singh and Gautam (2004) and Khan et al. (2009), have conducted studies into database usage. The current study is also an effort in this regard.

3.2.1 Research Framework

A student's satisfaction with online research databases can be understood in view of the TSM. Developed by Islam (2014), this model incorporates two essential motivational characteristics – computer self-efficacy and satisfaction – into the original technology acceptance model (TAM), developed by Davis et al. (1989). The TAM was derived from the theory of reasoned action (TRA). The TRA concentrates on the dimensions of consciously anticipated behaviours (Fishbein & Ajzen, 1975), which is a behavioural theory and popular in the business context. The TRA holds that belief can influence attitude, leading to intention and continuing to generate behaviour (Islam, 2016). The TAM asserts that beliefs about perceived ease of use and usefulness are consistently the key elements of information system acceptance in organisations. The literature suggests that the TAM has been frequently applied in library and information science, education, information systems, tourism, business and ICT. Islam (2014) identified the extensive potentiality of learners' CSE, which was the main contribution of social cognitive theory (SCT) where Bandura (1977) announced the idea of self-efficacy for the first time. In the case of technology satisfaction, Islam et al. (2015) argued that CSE should be included as an important dominant indicator in assessing students' satisfaction with using technology in higher education, though this area needs more investigation.

Forming the TSM through combining SCT and TAM was an innovative idea. Literature reveals that researchers have recently applied the TSM in studies in Malaysia, China and Pakistan and published their articles in the *Electronic Library, Journal of Information Science, Education and Information Technologies* and *SAGE Open*, among others. Jiang et al. (2021) stated that the TSM is one of the most indispensable models that has been tested as effective in determining learners' satisfaction in educational contexts. Islam et al. (2020, p. 14) stated, "advanced statistical analyses of the results revealed that the TSM has 'strong predictive power' in gauging Chinese lecturers' satisfaction with ICT". They also claimed that their study "contributes to relevance in advancing knowledge in the field of ICT assessment. 'Satisfaction' is a complicated concept to define, and this helps provide a model for doing so". Du et al. (2022, p. 1) asserted,

> our new findings on the moderating effect of local (i.e., Chinese) and international (e.g., English) academic databases highlighted that the TSM has successfully estimated dual databases and produced insignificant, dissimilar results. This study could aid local and international educators, researchers, information science professionals, and others in measuring the perception of academic databases for learning and research. This research could also serve as a guideline for researchers and psychometricians in measuring innovative learning technologies using structural equation modelling and the Rasch model. This is the unique contribution of the present study, which concludes that local and international academic databases are almost equally important for postgraduate students at a research university in China.

Islam et al. (2018, p. 12) found, "The results of the investigation have expanded the existing body of knowledge on student SAT with wireless internet and contributed to a better understanding of the TSM in various ways". Moreover, "The TSM was shown to be a useful model to measure students' SAT in using wireless internet in a different culture" (p. 2). Islam et al. (2015) tested the efficacy of the TSM, and they stated that their "findings confirmed that the TSM is viable to examine the successful integration of online research databases among postgraduate students in higher education". Technological emergences have aided academic libraries in creating groundbreaking mechanisms to efficiently assimilate digital resources (Rafi et al., 2022). Digital library databases have a substantial influence on stimulating the research culture in tertiary education. The usage of databases makes it conceivable to comprehend the research productivity and intellectual growth of user information needs (Demir, 2020; Rafi et al., 2019). Digital libraries are always considered an important facilitator of scientific research in academic institutions. The databases of digital libraries are more and more important for conducting high-quality research. This researcher have stated that, in the information age, database resources can help researchers come up with new ideas, direct research and make the academic communication between faculty and students more productive (Rafi et al., 2020; Machimbidza & Mutula, 2020). Hence, this study assumes the worthiness of adapting the TSM for testing online research databases.

According to the TSM, three factors can influence technology satisfaction (SAT). The first factor concerns the user's perception about the benefits of using technology, known as perceived usefulness (PU). The second factor is how easy users perceive the use of technology, referred to as perceived ease of use (PEU). The third factor is computer self-efficacy (CSE), which concerns the extent to which a person believes that their ability will help them succeed in what they are doing. It is also an important predictor of students' satisfaction with online databases. These three factors determine the extent of user's satisfaction with technology in higher education. In this study, satisfaction refers to the extent to which the postgraduate students find the use of technology consistent with their existing values, needs and experiences. The TSM contains one exogenous variable (CSE), two mediated variables (PEU and PU) and one endogenous variable (SAT).

The TAM emphasises that usefulness and ease of use are the key determinants of information technology and information system adoption in any organisation. The TAM stresses that only these two determinants form the basis of attitude towards a particular system, which, in turn, determine the intention of use and then generate the actual usage behaviour. In their original model (i.e., TAM), Davis et al. (1989) did not include the two essential motivation attributes, namely computer self-efficacy and satisfaction; these were subsequently added by Islam (2014) to aid a better understanding of TAM. However, Islam (2014) did not test the direct relationship between PEU and PU, the indirect relationship between CSE and PU, followed by PEU and SAT mediated by PEU and PU. The present study has factored in these relationships within the TSM, as indicated in Figure 3.1.

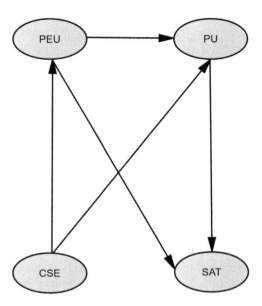

Figure 3.1 Technology satisfaction model (Islam, 2014).

Note: satisfaction (SAT), computer self-efficacy (CSE), perceived ease of use (PEU), and perceived usefulness (PU).

3.2.1.1 Research Hypotheses

This section presents a brief review of the literature pertaining to the factors impacting students' satisfaction with online research databases. On the basis of this review, nine hypotheses are presented to explain the factors influencing CUI students' satisfaction with online research databases. These hypotheses clarify the relationships among all the constructs under study, including computer self-efficacy, perceived usefulness, perceived ease of use and satisfaction in using the databases.

In a validation study of the TSM conducted by Islam (2014), the findings disclosed that computer self-efficacy is the most important predictor of students' satisfaction, which is directly related to the perceived usefulness of the wireless internet. The findings of Suki (2016) showed that the inclusion of self-efficacy in a model aimed at expanding knowledge gives useful insights about the essential dimensions that influence library patrons' intention to use public computing facilities. Ayoku and Okafor (2015) claimed that many users still have limited knowledge of databases and, therefore, self-efficacy should be the turning point to improve their knowledge. According to Aharony (2015), self-efficacy depends on self-views concerning dissimilar behaviours, which is considered to be condition-specific. In other words, Chang et al. (2016) have stated that seeking information and the quality of the attained results significantly affect people's views of their self-efficacy. Therefore, the authors of this paper believe that, since the concept of using online research databases for research and learning purposes is still in its infancy in Pakistan, prospective postgraduate students are most likely to face difficulties affecting user satisfaction. Hence, computer self-efficacy would play a vital role in database satisfaction. Based on empirical studies in students' use of online databases, researchers confirmed that an individual's sense of PU and PEU of the computer has a direct relationship with CSE (Du et al., 2022; Chen et al., 2020; Islam et al., 2015). Recently, a few studies have examined the association between computer self-efficacy and usefulness in measuring information technology (Islam et al., 2020; Bin et al., 2020; Islam et al., 2019; Shittu et al., 2016; Islam et al., 2018; Islam, 2016; Islam, 2014), online learning (Jiang et al., 2021, 2022) and computer simulation adoption (Liu & Huang, 2015).

While online research databases have been found to be effective in fostering research and enhancing the learning environment, only a few empirical studies have explored the association between computer self-efficacy and the perceived usefulness of using such facilities. Islam's (2014) study, for example, argues that computer self-efficacy has a significant direct impact on the perceived ease of use of the wireless internet. In a study involving sophomore students' use of a computer simulation tool, Liu and Huang (2015) found that self-efficacy is very beneficial for enhancing the perceived usefulness and ease of use. In their study, they also found that once learners achieve higher self-efficacy, they acquire a certain degree of self-confidence necessary for using a computer simulation tool. Li et al. (2022) modified the unified theory of acceptance and use of technology (UTAUT) model by adding three sub-components of self-efficacy and showed that self-efficacy influences effort and performance expectancies of mobile learning. Moreover, Chiu (2017) has suggested that in-service school teachers' computer self-efficacy is directly related to the perceived ease of use and usefulness of electronic textbooks. In the light of these findings, it is important to assess students' beliefs about their computer self-efficacy and their impact on the perceived ease of use and usefulness of online research databases. These observations, therefore, led us to hypothesise that in this study:

H1: Postgraduate students' computer self-efficacy has a direct impact on the perceived usefulness of online research databases.

H2: Postgraduate students' computer self-efficacy has a direct impact on the perceived ease of use of online research databases.

Studies related to technology adoption or acceptance have been continuously measured by two influential components, perceived usefulness and ease of use, as proposed by Davis (1989). They are valid for examining numerous technological facilities (Chiu, 2017; Huang, 2015; Huang, 2017; Liu & Huang, 2015; Mohammed et al., 2017; Teo & Zhou, 2014; Yan et al., 2016). Furthermore, the TSM has shown that perceived usefulness and ease of use are directly

82 *Applying the Rasch Model and Structural Equation Modeling to Higher Education*

associated with students' satisfaction in using the wireless internet technology (Islam, 2014; Islam, 2017; Islam et al., 2018), ICT facilities (Islam, 2016; Islam et al., 2019; Bin et al., 2020; Islam et al., 2020), online research databases (Du et al., 2022; Chen et al., 2020; Islam et al., 2015) and e-learning (Jiang et al., 2022; Jiang et al., 2021). Similarly, Masrek and Gaskin (2016) found that users' satisfaction was affected by the perceived ease of use of the web digital libraries. However, Yuan et al. (2016) and Joo and Choi (2016) exhibited the impact of perceived usefulness on user satisfaction with mobile banking and library resources. All of these results show that postgraduate students' satisfaction with online research databases in terms of perceived usefulness and ease of use is still underdeveloped. Thus, the author hypothesises:

H3: Perceived usefulness has a direct impact on postgraduate students' satisfaction with using online research databases.

H4: Perceived ease of use has a direct impact on postgraduate students' satisfaction with using online research databases.

The direct association between perceived ease of use and usefulness was established in the late 1980s (Davis et al., 1989; Chiu, 2017; Yuan et al., 2016). This researcher have also confirmed the validity of this kind of relationship in testing information systems (Davis et al., 1989), mobile learning (Poong et al., 2017), electronic textbooks (Chiu, 2017), digital libraries (Chen et al., 2016), cloud services (Huang, 2017), information technology (Teo & Zhou, 2014) and mobile banking (Yuan et al., 2016). In addition, several studies have claimed that learners' satisfaction is defined by the perceived ease of use (Islam, 2014; Islam et al., 2015; Masrek & Gaskin, 2016) and usefulness (Yuan et al., 2016; Islam et al., 2015; Islam, 2014). However, there is no evidence that the usefulness of online research databases can mediate the association between perceived ease of use and satisfaction. Hence, the author hypothesises:

H5: For postgraduate students, perceived ease of use has a direct impact on the perceived usefulness of online research databases.

H9: For postgraduate students, perceived ease of use has an indirect impact on their satisfaction mediated by the perceived usefulness of online research databases.

This researchers of this article believe that it is very common that learners' computer self-efficacy can determine their adoption or acceptance of new technological services, and that there is a significant association between computer self-efficacy, perceived ease of use and usefulness. For instance, Chiu (2017) has asserted that for in-service teachers, ease of use and usefulness of electronic textbooks depend on the teachers' computer self-efficacy. Notwithstanding this assertion, few studies have argued that learners' satisfaction can be influenced by their beliefs in their computer ability, mediated by the benefit and ease of use of databases. Therefore, the author hypothesises:

H6: Postgraduate students' computer self-efficacy has an indirect impact on their satisfaction mediated by the perceived usefulness of online research databases.

H7: Postgraduate students' computer self-efficacy has an indirect impact on their satisfaction mediated by the perceived ease of use of online research databases.

H8: Postgraduate students' computer self-efficacy has an indirect impact on the perceived usefulness mediated by the perceived ease of use of online research databases.

3.3 Methodology

This study intends to determine the extent to which the technology satisfaction model (TSM) validly measures the postgraduate students' satisfaction with online research databases in an academic library. It further explores the relationships among the antecedents within the model. To measure the components of the TSM and the concerns of prospective respondents, a survey was conducted. A total of 300 postgraduate students from six faculties of COMSATS University Islamabad were invited to participate in the study through a stratified random sampling procedure. This researchers chose to concentrate on postgraduate students because they have the most interactions with online databases for their research through the library website of CUI. The judgment sampling adequacy of this study was performed depending on psychometricians who stated that each item of the questionnaire should be completed by at least five respondents (Hair et al., 2010). According to the aforementioned recommendation, the sample size of this study should be at least 165 respondents because the questionnaire contains 33 items. Nevertheless, this research distributed 300 questionnaires to ensure the sampling adequacy. On the other hand, Hu and Bentler (1999) claimed that for structural equation modeling, an adequate sample size is larger than 250. This study received 263 completed student responses, which constitutes more than a sufficient sample size (Hair et al., 2010; Hu & Bentler, 1999). A total of 263 students responded to the survey, achieving a response rate of 87.66%. The data were statistically analysed through the SPSS software version 22. The majority of the respondents (162; 61.59%) were male and 101 (38.40%) were female; 107 students (41.5%) were between the ages of 21 and 25. Most of the respondents (235; 89.35%) were at the master's level, whereas 28 (10.64%) were at the PhD level. The questionnaire was divided into five parts and consisted of 33 items, measuring the four constructs pertaining to postgraduate students' satisfaction of the COMSATS online research databases: perceived ease of use, perceived usefulness, computer self-efficacy and satisfaction. The questionnaire was adapted from a previous

84 *Applying the Rasch Model and Structural Equation Modeling to Higher Education*

Table 3.1 Components Measured in the Questionnaire

Components	Likert Scale	No. of Items
Computer Self-Efficacy	1 to 5 (strongly disagree → strongly agree)	8
Perceived Ease of Use		10
Perceived Usefulness		10
Satisfaction	1 to 5 (very unsatisfied → very satisfied)	5
Total		*33*

study by Islam (2011). To measure these four constructs, this researchers used a five-point Likert scale as shown in Table 3.1.

The scale's reliability and validity were tested using Rasch analysis through Winsteps version 3.49. The Rasch (1960) model has been widely used by researchers to develop and validate the instrument. The instrument was estimated based on several Rasch outputs, namely summary statistics, item polarity map, item fit order, item map and principal components. The summary statistics and item polarity map provide information on the reliability of the instrument, while the remaining outputs, that is item fit order, item map and principal components, offer detailed information on the validity of the scale. The output of the summary statistics showed that item reliability and person reliability were 0.92 and 0.93. Moreover, the item separation and person separation were 3.37 and 3.62 as shown in Figure 3.2. The item polarity map indicates that a total of 33 items were measured in the same direction, and their point measure correlations (PTMEA CORR.) were greater than .43.

Concerning the validity of the items, the Rasch output of item fit order suggested that two items were outside the range of infit and outfit mean square values, and they belonged to two different dimensions, namely perceived usefulness (PU4) and perceived ease of use (PEU6). For the rating scale, infit and outfit mean square values should be between 0.5 and 1.5 (Bond & Fox, 2001). These misfitting items were excluded from the original pool of 33 items for further analysis.

The fundamental concept of the Rasch model is that it is more expected for all persons to respond to the easy items correctly than complicated items, and it is more expected for all items to be answered by persons who have higher ability than lower ability (Bond & Fox, 2001). The results of the item map recommended that the majority of the postgraduate students were satisfied with the online research databases, although students' ability was found to be higher than the items' difficulty as shown in Figure 3.3. Therefore, the inclusion of some difficult items in future studies is required in order to improve students' satisfaction. Overall, the results of this study supported the theoretical structure of the psychometric properties of computer self-efficacy, satisfaction, and the perceived ease of use and usefulness of research databases. Thirty-one items fit the model and have a good scale (see Table 3.2). The proportion of variance explained by the measures was 52.1, which indicated that the items were able to endorse the students' satisfaction with the databases.

Assessing Online Research Databases in HE Using the TSM 85

```
Figure     Online Research Database                ZOU999ws.txt Sep  8  0:02 2018
INPUT: 263 PERSONS, 33 ITEMS  MEASURED: 263 PERSONS, 33 ITEMS, 5 CATS        3.49
--------------------------------------------------------------------------------

       SUMMARY OF 243 MEASURED (NON-EXTREME) PERSONS
+------------------------------------------------------------------------------+
|           RAW                          MODEL         INFIT        OUTFIT      |
|          SCORE      COUNT    MEASURE    ERROR     MNSQ   ZSTD   MNSQ    ZSTD   |
|------------------------------------------------------------------------------|
| MEAN     136.8       33.0       2.17      .32     1.05    -.1   1.02    -.2    |
| S.D.      15.6         .0       1.38      .10      .65    2.4    .63    2.4    |
| MAX.     164.0       33.0       6.49     1.01     4.10    7.9   3.99    7.8    |
| MIN.      79.0       33.0      -1.00      .19      .06   -7.6    .06   -7.2    |
|------------------------------------------------------------------------------|
| REAL RMSE    .37  ADJ.SD    1.33  SEPARATION  3.62  PERSON RELIABILITY   .93   |
|MODEL RMSE    .33  ADJ.SD    1.34  SEPARATION  4.07  PERSON RELIABILITY   .94   |
| S.E. OF PERSON MEAN = .09                                                     |
+------------------------------------------------------------------------------+

  MAXIMUM EXTREME SCORE:     9 PERSONS
  MINIMUM EXTREME SCORE:     1 PERSONS


       SUMMARY OF 33 MEASURED (NON-EXTREME) ITEMS
+------------------------------------------------------------------------------+
|           RAW                          MODEL         INFIT        OUTFIT      |
|          SCORE      COUNT    MEASURE    ERROR     MNSQ   ZSTD   MNSQ    ZSTD   |
|------------------------------------------------------------------------------|
| MEAN    1007.2      243.0        .00      .11      .99    -.2   1.02     .0    |
| S.D.      36.0         .0        .39      .01      .27    2.4    .34    2.7    |
| MAX.    1076.0      243.0       1.05      .12     1.92    7.0   2.31    9.2    |
| MIN.     900.0      243.0       -.87      .09      .70   -3.1    .64   -3.2    |
|------------------------------------------------------------------------------|
| REAL RMSE    .11  ADJ.SD     .38  SEPARATION  3.37  ITEM    RELIABILITY   .92  |
|MODEL RMSE    .11  ADJ.SD     .38  SEPARATION  3.53  ITEM    RELIABILITY   .93  |
| S.E. OF ITEM MEAN = .07                                                       |
+------------------------------------------------------------------------------+
```

Figure 3.2 The summary statistics.

A confirmatory factor analysis (CFA) was performed by the 31 valid items to confirm the factor structure of the TSM, using the results from the Rasch model. To investigate the structural associations among the various facets of the TSM, the analysis used the utmost probability estimation to generate the estimates of the measurement and structural models. In other words, the models were assessed for the goodness-of-fit of each model (Hu & Bentler, 1999), including chi-square/degree of freedom ($\chi^2/df \le 3$), root mean square error of approximation (RMSEA $\le .08/.1$), Tucker–Lewis index (TLI $\ge .90$) and comparative fit index (CFI $\ge .90$). These comprise the key elements of fit statistics. Based on these fit indices, we were able to decide whether the models fit the data well for further estimation.

3.4 Results

The four-factor measurement model of perceived ease of use, satisfaction, perceived usefulness and computer self-efficacy was validated by applying

86 *Applying the Rasch Model and Structural Equation Modeling to Higher Education*

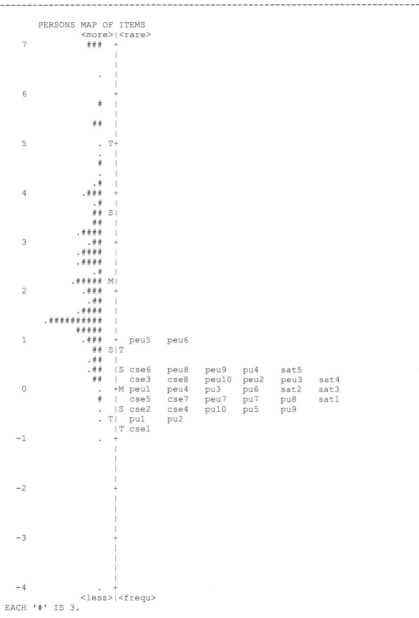

Figure 3.3 Item map.

Table 3.2 The 31 Valid Items

Constructs	Valid Items		α
PEU	PEU1	I find the university online research databases easy to use.	.885
	PEU2	I find it easy to access the university online research databases.	
	PEU3	It is easy for me to become skilful at using online research databases.	
	PEU4	It is easy for me to remember how to search journals/articles by using the online research databases.	
	PEU5	Interacting with the online research databases system requires minimal mental effort.	
	PEU7	I find it easy to get the online research databases to help me do my research.	
	PEU8	My interaction with the online research databases is within my comprehension.	
	PEU9	Interacting with the online research databases system is very stimulating for me.	
	PEU10	I find it easy to choose articles/journals of different categories, for example, education, engineering and business.	
PU	PU1	Using the online research databases enables me to get research materials.	.926
	PU2	Using the online research databases helps me write my thesis/journal articles.	
	PU3	The online research database addresses my thesis-related needs.	
	PU5	Using the online research databases system saves me time.	
	PU6	Using the online research databases system allows me to accomplish more work than would otherwise be possible.	
	PU7	Using the online research databases improves the quality of my work.	
	PU8	Using the online research databases enhances my knowledge and research skills.	
	PU9	Using the online research databases increases my research productivity.	
	PU10	Using the online research databases provides me with the latest information on particular areas of research.	
CSE	CSE1	I am able to use the online research databases.	.901
	CSE2	I can download research materials from the online research databases.	
	CSE3	I can navigate my way through the online research databases.	
	CSE4	I can use the online research databases to enhance the quality of my research.	
	CSE5	I can save and print journals/articles from the online research databases.	
	CSE6	I have the ability to e-mail journals/articles from online research databases.	
	CSE7	I have the knowledge and skills required to benefit from using the online research databases.	
	CSE8	I can access the online research databases from the university website.	
SAT	SAT1	Overall, I am satisfied with the ease of completing my task by using online research databases.	.880
	SAT2	The university online research databases service has greatly affected the way I search for information and conduct my research.	
	SAT3	Providing the online research databases is an indispensable and satisfactory service provided by the university.	
	SAT4	Overall, I am satisfied with the amount of time it takes to complete my task by using online research databases.	
	SAT5	I am satisfied with the structure of accessible information (available as categories of a research domain, or by the date of issue – for journals in particular – or as full-texts or abstracts of theses and dissertations) of the online databases.	

CFA to analyse and confirm the factor structure of the TSM based on the results from the Rasch analysis. This measurement model also explained whether postgraduate students' satisfaction was interrelated with their abilities, ease of use and usefulness of databases. To test the relationships among the factors, the preliminary CFA model was performed on 31 items, using maximum likelihood estimation techniques, which explain the four-factor measurement model. With the exception of the CFI ($\leq$ .90) and TLI ($\leq$ .90), which were less suitable, the findings of the measurement model attested that the set of fit statistics showed that the data were well fit by the measurement model: χ^2 = 1261.514; df = 428; p = 0.000; and RMSEA = 0.088 ($\leq$ .1). The four-factor measurement model was tested using the AMOS programme, which exhibited significant interrelationships between the facets, and all the factor loadings were greater than .59; however, a few items indicated high modification indices and contradicted the convergent and discriminant validity. Hence, the measurement model required modification because of further improvement of the fit statistics and confirmation of the convergent and discriminant validity.

To improve the fit statistics and confirm the convergent and discriminant validity of the four-factor measurement model, the authors excluded a number of indicators (PEU1, PEU2, PEU5, CSE1, CSE4, CSE7, CSE8, SAT2, PU1, PU3 and PU5), one at a time and based on their modification indices, with the largest being dropped first from all the components of the TSM. The modified four-factor measurement model suggested an exceptional goodness-of-fit of the model: χ^2 = 393.389; df = 164; p = 0.000; RMSEA = 0.075 ($\leq$ .08); CFI = 0.923($\geq$ .90); and TLI = 0.911 ($\geq$ .90). The modified measurement model contained a total of 20 valid parameters from the original pool of 31 indicators, where all the factor loadings of the items ranged from .65 to .87. The authors also estimated the convergent and discriminant validity by calculating composite reliability and variance, extracted using the formula of Fornell and Larcker (1981).

Table 3.3 shows that all the scores of composite reliability (CR) were greater than .845, which confirms that the indicators adopted for each facet were following their respective dimensions and were reliable. The t-values of parameters were statistically significant at the .05 level, showing that all the indicators within each dimension were significantly correlated with each other and, thus, exhibited convergent validity. The authors computed each facet's average variance extracted (AVE), which also supported convergent validity, since they all exceeded 0.50 (Hair et al., 2010), ranging from 0.502 to 0.612. The modified four-factor measurement model showed that the interrelationships, namely between CSE and PEU (β = 0.70, p < .001), CSE and PU (β = 0.71, p < .001), CSE and SAT (β = 0.77, p < .001), PEU and SAT (β = 0.75, p < .001), PU and SAT (β = 0.77, p < .001) and PEU and PU (β = 0.75, p < .001), were statistically significant and the correlations between the facets were smaller than one, supporting the satisfactory discriminant validity of the psychometric properties.

Assessing Online Research Databases in HE Using the TSM 89

Table 3.3 The Loadings for the Revised Four-Factor Measurement Model

Factors/Items		Loadings	M	SD	α	CR	AVE
Perceived Ease of Use (PEU)					0.857	0.858	0.502
PEU3	It is easy for me to become skilful at using online research databases.	.72	4.09	.802			
PEU4	It is easy for me to remember how to search journals/articles by using the online research databases.	.74	4.17	.803			
PEU7	I find it easy to get the online research databases to help me do my research.	.75	4.21	.840			
PEU8	My interaction with the online research database is within my comprehension.	.71	4.04	.760			
PEU9	Interacting with the online database system is very stimulating for me.	.69	4.02	.811			
PEU10	I find it easy to choose articles/ journals of different categories (e.g., education, engineering and business).	.65	4.13	.810			
Computer Self-Efficacy (CSE)					0.844	0.845	0.578
CSE2	I can download research materials from the online research databases.	.79	4.30	.853			
CSE3	I can navigate my way through the online research databases.	.79	4.12	.738			
CSE5	I can save and print journals/articles from the online research databases.	.74	4.22	.814			
CSE6	I have the ability to e-mail journals/ articles from online research databases.	.71	4.03	.875			
Satisfaction (SAT)					0.848	0.851	0.590
SAT1	Overall, I am satisfied with the ease of completing my task.	.79	4.25	.711			
SAT3	Providing the online research databases is an indispensable and satisfactory service provided by the university.	.73	4.19	.749			
SAT4	Overall, I am satisfied with the amount of time it takes to complete my task.	.83	4.11	.863			
SAT5	I am satisfied with the structure of accessible information (available as categories of a research domain, or by the date of issue – for journals in particular – or as full-texts or abstracts of theses and dissertations) of the online databases.	.72	4.03	.840			
Perceived Usefulness (PU)					0.900	0.904	0.612
PU2	Using the online database helps me write my thesis/journal articles.	.70	4.33	.696			
PU6	Using the online database system allows me to accomplish more work than would otherwise be possible.	.71	4.18	.776			

(Continued)

90 *Applying the Rasch Model and Structural Equation Modeling to Higher Education*

Table 3.3 (Continued)

Factors/Items		Loadings	M	SD	α	CR	AVE
PU7	Using the online database improves the quality of my work.	.76	4.22	.693			
PU8	Using the online database enhances my knowledge and research skills.	.83	4.26	.682			
PU9	Using the online databases increases my research productivity.	.87	4.28	.698			
PU10	Using the online databases provides me with the latest information on particular areas of research.	.81	4.28	.739			

3.5 Estimating the Structural Model

In this research, the TSM was considered a structural model, which was used to examine all the proposed hypotheses. First, the validity of the TSM was claimed based on four common model-fit measures such as χ^2/df, RMSEA, CFI and TLI. These fit indices confirmed that the data were well fit by the TSM: $\chi^2 = 411.782$; $df = 165$; $p = 0.000$; RMSEA = 0.077 ($\leq .08$); CFI = 0.917 ($\geq .90$); and TLI = 0.905 ($\geq .90$). As indicated in Figure 3.4, all of the loadings of the TSM were statistically significant and greater than .65. All the path coefficients between the facets of the TSM were analysed using standardised estimates. As observed in the TSM, which was achieved as a result of the present study, the postgraduate students' computer self-efficacy was identified to have a direct impact on the perceived usefulness ($\beta = 0.39$, $p < 0.001$) and ease of use ($\beta = 0.72$, $p < 0.001$) of online research databases. These findings indicate the accuracy of the first two hypotheses (*H1* and *H2*). Subsequently, perceived usefulness ($\beta = 0.48$, $p < 0.001$) and ease of use ($\beta = 0.40$, $p < 0.001$) were identified as having a direct impact on postgraduate students' satisfaction with online research databases for research and learning purposes. These relationships attest to the accuracy of *H3* and *H4*. With regard to *H5*, the TSM showed that for the learners, the perceived ease of use had a direct impact on the perceived usefulness ($\beta = 0.47$, $p < 0.001$) of online research databases, thereby supporting this hypothesis. Regarding the mediating relationships, the Sobel test confirmed that the postgraduate learners' computer self-efficacy had an indirect impact on their satisfaction, mediated by the perceived usefulness (chi-square, $\chi^2 = 3.494$; $p = 0.000$) and ease of use (chi-square, $\chi^2 = 3.701$; $p = 0.000$) of online research databases. These results validate *H6* and *H7* (Sobel, 1982). However, computer self-efficacy was also identified to have an indirect impact on perceived usefulness mediated by perceived ease of use (chi-square, $\chi^2 = 4.662$; $p = 0.000$), thereby validating *H8*.

The TSM also detected that for the learners, perceived ease of use had an indirect impact on their satisfaction, mediated by the perceived usefulness (chi-square, $\chi^2 = 3.451$; $p = 0.000$) of online research databases. This finding verifies *H9*. In addition, this research illustrated that computer self-efficacy was

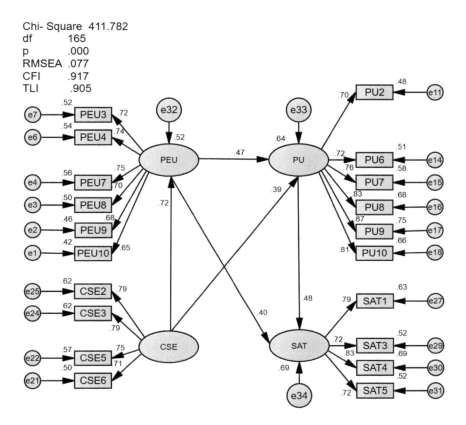

Figure 3.4 The technology satisfaction model.

the most powerful facet of the TSM. To determine the extrapolative power of the TSM, this study observed the explained variance (R^2) of the mediating and endogenous variables. The R^2 values of the model showed that 69% of the variance in the postgraduate students' satisfaction was explained by computer self-efficacy, perceived ease of use and perceived usefulness, which confirmed that the TSM had a strong predictive power and was viable for measuring endogenous variable. Similarly, the model also indicated that 64% and 52% of the variances in learners' PU and PEU were explained by CSE. Finally, Table 3.4 summarises the hypotheses and the critical ratios (CR) together with the total standardised direct and indirect effect sizes of the facets of the TSM.

3.6 Discussion

Drawing results from the existing literature about online research databases, we found that TSM was the core model. The findings of the TSM have expanded the existing body of knowledge on postgraduate students' satisfaction with

92 *Applying the Rasch Model and Structural Equation Modeling to Higher Education*

Table 3.4 A Summary of the Hypotheses and Total Standardised Direct and Indirect Effect Sizes

Hypotheses	Linkages	Path Coefficients (β)	CR (p)	Effect Size	Results
Direct Effects					
H1	CSE→PU	0.39	4.312 (0.000)	0.385	Accepted
H2	CSE→PEU	0.72	7.846 (0.000)	0.723	Accepted
H3	PU→SAT	0.48	5.221 (0.000)	0.480	Accepted
H4	PEU→SAT	0.40	4.364 (0.000)	0.405	Accepted
H5	PEU→PU	0.47	5.023 (0.000)	0.475	Accepted
Indirect Effects			$\chi 2\ (p)$		
H6	CSE→PU→SAT	0.187 (≥ 0.080)	3.494 (0.000)	0.642	Accepted
H7	CSE→PEU→SAT	0.288 (≥ 0.080)	3.701 (0.000)		Accepted
H8	CSE→PEU→PU	0.338 (≥ 0.080)	4.662 (0.000)	0.343	Accepted
H9	PEU→PU→SAT	0.225 (≥ 0.080)	3.451 (0.000)	0.228	Accepted

online research databases in several ways. Using the TSM, nine hypotheses were constructed to show the relationships among the exogenous (computer self-efficacy), endogenous (satisfaction) and mediating (perceived usefulness and ease of use) variables. The empirical findings from the structural equation modeling (SEM) analyses suggest that the TSM is a better fit, indicating that the four antecedents are suitable determinants for assessing postgraduate students' satisfaction with online research databases in Pakistan.

The findings from the TSM support *H1* and *H2*, illustrating that the postgraduate students' computer self-efficacy has a direct impact on the perceived usefulness and ease of use of online research databases. This implies that the benefits and ease of use of online research databases depend on the postgraduate students' computer self-efficacy. Liu and Huang (2015) have claimed that self-efficacy is very beneficial for enhancing usefulness and ease of use. Once students have achieved higher self-efficacy, they acquire a certain degree of self-confidence necessary for using computer tools. Therefore, libraries should ensure that postgraduate students not only have the required IT skills to use databases, but also feel that online research databases are easy to use and benefit their learning and research activities. Students' abilities in using databases can be improved through ongoing training programmes. To ensure the quality of the training programmes, libraries are advised to conduct evaluation research to identify their strengths and weaknesses. Moreover, libraries should assess trainers' performance in facilitating the training programmes.

The results of this research are notably congruent with a recent TSM-based study on the wireless internet (Islam, 2014). In subsequent studies, other researchers have found an association between computer self-efficacy and usefulness in measuring computer simulation adoption (Liu & Huang, 2015) and information technology (Bin et al., 2020; Islam et al., 2019; Shittu et al., 2016; Islam, 2016; Islam et al., 2018; Islam et al., 2020). This researcher have argued that school teachers' computer self-efficacy is directly related to the perceived ease of use and usefulness of electronic textbooks (Chiu, 2017), online research databases (Du et al., 2022; Chen et al., 2020; Islam et al., 2015) and e-learning (Jiang et al., 2021, 2022).

The results of this study pertaining to the impacts of perceived usefulness (*H3*) and ease of use (*H4*) of online research databases on the postgraduate students' satisfaction with using these facilities echo the results reported by Islam (2014), who developed and validated the TSM for assessing the wireless internet; the findings of this study also echo those reported by other researchers, such as Yuan et al. (2016), Joo and Choi (2016) and Masrek and Gaskin (2016), who conducted studies on mobile banking, library resources and web digital libraries, respectively. However, two extrinsic motivation components of the TAM (Davis et al., 1989), namely perceived usefulness and ease of use, have been frequently incorporated by researchers investigating technology adoption or acceptance studies (Chiu, 2017; Huang, 2015; Huang, 2017; Liu & Huang, 2015; Mohammed et al., 2017; Teo & Zhou, 2014; Yan et al., 2016). Interestingly, the results of this research in Pakistan conclude that the postgraduate students' satisfaction depends on the perceived ease of use and benefits of research databases. This suggests that once students feel that databases are easy to use and useful for their learning and research, they will gain great satisfaction from using them. Therefore, to increase students' level of satisfaction, libraries should have an ongoing evaluation programme with a view to improving the quality of their services.

In addition, the findings of this study demonstrate a direct association between perceived ease of use and perceived usefulness of online research databases (*H5*), echoing findings from studies related to technology acceptance (Chen et al., 2016; Chiu, 2017; Davis et al., 1989; Poong et al., 2017; Teo & Zhou, 2014); the findings of this research also illustrate an indirect association between perceived ease of use and satisfaction through perceived usefulness (*H9*). These can be accepted as a contribution to the TSM, although these two hypotheses have not been tested previously. These associations confirm that easy access to databases is one of the vital issues that can help postgraduate students feel the benefits of this service, leading to enhanced levels of satisfaction. Hence, libraries are advised to ensure that postgraduate students can easily access the required information through their databases at any time, anywhere. This could be done by providing a high-speed internet service along with well-equipped computer labs and multiple access points.

Finally, the findings show that the postgraduate students' computer self-efficacy has an indirect impact on their satisfaction through perceived usefulness (*H6*) and ease of use (*H7*) of online research databases. The results of *H6* and *H7* are similar to that of Islam (2014), who conducted research on the wireless internet. Furthermore, the inclusion of *H8*, a new hypothesis, in the TSM illustrates that the postgraduate students' computer self-efficacy has an indirect impact on the perceived usefulness through perceived ease of use of online research databases.

In the light of the findings of this study, it is evident that postgraduate students' satisfaction with online research databases does not depend on extrinsic motivation attributes alone, that is perceived usefulness and ease of use, but also on their belief in their own ability to use it. This suggests that students' belief in their ability to use databases and acknowledgement of their value for them are most likely to affect their satisfaction. Librarians should ensure that postgraduate students have the requisite technological knowledge and skills before offering them extensive research databases. Hence, students' ability (computer self-efficacy) to use databases can be recognised as a distinct predictor of the TSM, which is an important intrinsic motivation factor in technology satisfaction. Students' computer self-efficacy alone can directly and indirectly impact the benefits, ease of use and satisfaction with online research databases.

This study has some important practical implications for both librarians and students. First, this study is significant because, by validating the TSM, it offers academic librarians a tool to predict students' satisfaction with research databases. Librarians in any part of the world can apply this model to assess user satisfaction with technology. Second, this study highlights the importance of students' self-efficacy in relation to their use of research databases. This is particularly important for libraries that strive to meet their users' expectations and contribute to the advancement of the educational goals of their institutions, with a view to improving students' retention and their academic growth. In such cases, library professionals need to concentrate on conducting regular digital information literacy programmes. Libraries should offer the digital information literacy programmes to the newly enrolled students at the start of the academic year. This will certainly help these students improve their IT skills, familiarise themselves with the research databases and gain maximum benefits from these research resources. This will also help the new students manage their academic and research assignments and fulfil their academic tasks more efficiently. The library and information science (LIS) professionals are also advised to incorporate digital information literacy programmes into the academic curriculum.

3.7 Conclusion

The application of the TSM was found to be appropriate for determining the Pakistani postgraduate students' satisfaction with online research databases to perform their academic and research activities. This empirical study illustrates that students' satisfaction with online research databases is directly and indirectly affected by their self-efficacy and the ease of use and benefits of using these resources. The findings suggest that to increase student satisfaction, service providers need to assure that databases are easy to use, and that their students have the required knowledge and skills to benefit from these databases.

Evidence from the TSM indicates that the direct and indirect impact of the postgraduate students' computer self-efficacy on satisfaction, perceived ease of use and usefulness of online research databases is limited to the extent of their requisite skills in downloading research materials from the online research databases, navigating their way through the databases, e-mailing journals or articles from the databases, and saving and printing articles from research databases.

The impact of the perceived usefulness of databases on the postgraduate students' satisfaction is manifested in their ability to accomplish more research work, write their theses or journal articles, improve the quality of their work, enhance their knowledge and research skills, increase their research productivity and obtain the latest information on specific areas of research.

For postgraduate students, the impact of the perceived ease of use on usefulness and satisfaction in using online research databases is illustrated in terms of their easy access to become skilful at using databases, the ease in which they can remember how to search journal articles through the databases, the ease of doing research through databases, the minimal effort required to interact with research databases system, and the interaction being stimulating. It is also evident that students found it easy to select articles/journals of different categories using university research databases.

In addition, it is worth mentioning that the Rasch model has been frequently applied to develop and validate the instruments in education, business, psychology, health and social sciences. However, there is a lack of evidence to show that researchers have developed and validated instruments using the Rasch model, especially in the information and library science field. From the literature, it is evident that most of the researchers have traditionally developed or validated questionnaires using exploratory factory analysis (EFA), focusing only on items, while the Rasch model allows researchers to test the reliability and validity of the items *and* persons.

Additionally, the results of the SEM analysis confirmed that a substantial extent of variance on satisfaction (69%) was explained by computer self-efficacy, perceived usefulness and ease of use. This result provides support for the theoretical structure of the TSM and illustrates its importance in online research databases. In addition, this study illustrates that perceived usefulness and ease of use had a noteworthy impact on postgraduate students' satisfaction with online research databases. These predictors were directly associated with user satisfaction, and they played a vital role in assessing databases. For postgraduate students, the perceived ease of use also had a direct impact on the perceived usefulness of online research databases.

As the postgraduate students' satisfaction is determined by the three valid facets of the TSM, some other components might also be included, such as computer anxiety and service quality of online research databases. Caution needs to be exercised when extrapolating the findings of this study to other settings. Therefore, it would be useful to conduct more research into online research databases in other universities using the TSM. Future researchers may undertake a comparative research on lecturers *and* postgraduate students to establish a better understanding of user satisfaction with online research. Similarly, future studies may be performed cross-culturally to identify the role of culture in the TSM. Libraries are advised to consider the benefits of conducting follow-up focus groups and/or interviews with postgraduate students to learn about their challenges. Lastly, future researchers are recommended to consider the moderating variables to complement the four facets of the TSM, such as age, gender, level of study, nationality and faculty.

3.8 Acknowledgements

This work was supported by the Peak Discipline Construction Project of Education at East China Normal University and Fundamental Research Funds for the Central Universities (2020ECNU-HLYT035).

3.9 References

Aagaard, P., & Arguello, N. Z. (2015). Practical approaches to compliance for entrepreneurial uses of licensed databases in libraries. *Reference Services Review, 43*(3), 419–438.

Aharony, N. (2015). An exploratory study on factors affecting the adoption of cloud computing by information professionals. *The Electronic Library, 33*(2), 308–323.

Ahmed, S. M. Z. (2013). Use of electronic resources by the faculty members in diverse public universities in Bangladesh. *Electronic Library, 31*(3), 290–312.

Albitz, B. (2008). *Licensing and managing electronic resources*. Elsevier.

Ani, O. E., & Ahiauzu, B. (2008). Towards effective development of electronic information resources in Nigerian university libraries. *Library Management, 29*(6/7), 504–514.

Atakan, C., Atılgan, D., Bayram, Ö., & Arslantekin, S. (2008). An evaluation of the second survey on electronic databases usage at Ankara University Digital Library. *The Electronic Library, 26*(2), 249–259.

Ayoku, O. A., & Okafor, V. N. (2015). ICT skills acquisition and competencies of librarians: Implications for digital and electronic environment in Nigerian universities libraries. *The Electronic Library, 33*(3), 502–523.

Ayoo, P. O., & Lubega, J.T. (2014). A framework for e-learning resources sharing (FeLRS). *International Journal of Information and Education Technology, 4*(1), 112–119.

Bandura, A. (1977). Self-efficacy: Toward a unifying theory of behavioral change. *Psychological Review, 84*(2), 191–215.

Bates, J., Best, P., McQuilkin, J., & Taylor, B. (2017). Will web search engines replace bibliographic databases in the systematic identification of research? *The Journal of Academic Librarianship, 43*(1), 8–17.

Bin, E., Islam, A. Y. M. A., Gu, X., Spector, J. M., & Wang, F. (2020). A study of Chinese technical and vocational college teachers' adoption and gratification in new technologies. *British Journal of Educational Technology, 51*(6), 2359–2375. https://doi.org/10.1111/bjet.12915

Bond, T. G., & Fox, C. M. (2001). *Applying the Rasch model: Fundamental measurement in the human science*. Lawrence Erlbaum.

Brar, I. S. (2015, June 29–July 5). Digital information literacy among health sciences professionals: A case study of GGS Medical College, Faridkot, Punjab, India. In *Proceedings of informing science & IT education conference, Tampa, FL* (pp. 531–541). Informing Science Institute.

Byerley, S. L., Chambers, M.B & Thohira, M. (2007). Accessibility of web-based library databases: The vendors' perspectives in 2007. *Library Hi Tech, 25*(4), 509–527.

Chang, Y., Wong, S. F., & Park, M. C. (2016). A three-tier ICT access model for intention to participate online: A comparison of developed and developing countries. *Information Development, 32*(3), 226–242.

Chen, H., Islam, A. Y. M. A., Gu, X., Teo, T., & Peng, Z. (2020). Technology-enhanced learning and research using databases in higher education: The application of the ODAS model. *Educational Psychology, 40*(9), 1056–1075. https://doi.org/10.1080/01443410.2019.1614149

Chen, J. F., Chang, J. F., Kao, C. W., & Huang, Y. M. (2016). Integrating ISSM into TAM to enhance digital library services: A case study of the Taiwan digital meta-library. *The Electronic Library, 34*(1), 58–73.

Chiu, T. K. (2017). Introducing electronic textbooks as daily-use technology in schools: A top-down adoption process. *British Journal of Educational Technology, 48*(2), 524–537.

Collins, T. (2012). The current budget environment and its impact on libraries, publishers and vendors. *Journal of Library Administration, 52*(1), 18–35.

Coombs, K. A. (2005). Lessons learned from analyzing library database usage data. *Library Hi Tech, 23*(4), 598–609.

Davis, F. D. (1989). Perceived usefulness, perceived ease of use, and user acceptance of information technology. *MIS Quarterly, 13*(3), 319–340.

Davis, F. D., Bagozzi, R. P., & Warshaw, P. R. (1989). User acceptance of computer technology: A comparison of two theoretical models. *Management Science, 35*(8), 982–1003.

Demir, S. B. (2020). Scholarly databases under scrutiny. *Journal of Librarianship and Information Science, 52*(1), 150–160.

Dermody, K., & Majekodunmi, N. (2011). Online databases and the research experience for university students with print disabilities. *Library Hi Tech, 29*(1), 149–160.

Du, H., Chen, H., & Islam, A. Y. M. A. (2022). Students' perception of academic databases as recognition of learning and research during the COVID-19 pandemic. *Journal of Information Science*. https://journals.sagepub.com/doi/full/10.1177/01655515221118666

Dukić, D. (2013). Online databases as research support and the role of librarians in their promotion: The case of Croatia. *Library Collections, Acquisitions, and Technical Services, 37*(1–2), 56–65.

Dukić, D. (2014). Use and perceptions of online academic databases among Croatian University teachers and researchers. *Libri, 64*(2), 173–184.

Falk, H. (2005). State library databases on the internet. *The Electronic Library, 23*(4), 492–498.

Fishbein, M., & Ajzen, I. (1975). *Belief, attitude, intention and behavior: An introduction to theory and research*. Addison-Wesley.

Fornell, C., & Larcker, D. F. (1981). Evaluating structural equation models with unobservable variables and measurement error. *Journal of Marketing Research, 48*, 39–50.

Guruprasad, R., Marimuthu, P., & Nikam, K. (2012). Use patterns of core aerospace engineering e-databases, gateways and standards: A research survey of aerospace scientists and engineers of Bangalore. *International Journal of Applied Engineering and Technology, 2*(1), 67–91.

Hair, J. F., Black, W., Babin, B. J., Anderson, R. E., & Tatham, R. L. (2010). *Multivariate data analysis a global perception*. Pearson Education Inc.

Horwath, J. (2002). Evaluating opportunities for expanded information access: A study of the accessibility of four online databases. *Library Hi Tech, 20*(2), 199–206.

Hu, L. T., & Bentler, P. M. (1999). Cutoff criteria for fit indexes in covariance structure analysis: Conventional criteria versus new alternatives. *Structural Equation Modeling, 6*(1), 1–55.

Huang, T. K. (2015). Exploring the antecedents of screenshot-based interactions in the context of advanced computer software learning. *Computers & Education, 80*, 95–107.

Huang, Y. M. (2017). Exploring the intention to use cloud services in collaboration contexts among Taiwan's private vocational students. *Information Development*, *33*(1), 29–42.

Islam, A. Y. M. A. (2011). *Online database adoption and satisfaction model*. Lambert Academic Publishing.

Islam, A. Y. M. A. (2014). Validation of the technology satisfaction model (TSM) developed in higher education: The application of structural equation modeling. *International Journal of Technology and Human Interaction*, *10*(3), 44–57.

Islam, A. Y. M. A. (2016). Development and validation of the technology adoption and gratification (TAG) model in higher education: A cross-cultural study between Malaysia and China. *International Journal of Technology and Human Interaction*, *12*(3), 78–105.

Islam, A. Y. M. A. (2017). Technology satisfaction in an academic context: Moderating effect of gender. In A. Mesquita (Ed.), *Research paradigms and contemporary perspectives on human-technology interaction* (pp. 187–211). IGI Global. https://doi.org/10.4018/978-1-5225-1868-6.ch009

Islam, A. Y. M. A., Ahmad, K., Rafi, M., & JianMing, Z. (2021). Performance-based evaluation of academic libraries in the big data era. *Journal of Information Science*, *47*(4), 458–471. https://doi.org/10.1177/0165551520918516

Islam, A. Y. M. A., Gu, X., Crook, C., & Spector, J. M. (2020). Assessment of ICT in tertiary education applying structural equation modeling and Rasch model. *SAGE Open*, *10*(4), 1–17. https://doi.org/10.1177/2158244020975409

Islam, A. Y. M. A., Leng, C. H., & Singh, D. (2015). Efficacy of the technology satisfaction model (TSM): An empirical study. *International Journal of Technology and Human Interaction*, *11*(2), 45–60.

Islam, A. Y. M. A., Mok, M. M. C, Gu, X., Spector, J. M., & Leng, C. H. (2019). ICT in higher education: An exploration of practices in Malaysian universities. *IEEE Access*, *7*(1), 16892–16908. https://doi.org/10.1109/ACCESS.2019.2895879

Islam, A. Y. M. A., Mok, M. M. C, Xiuxiu, Q., & Leng, C. H. (2018). Factors influencing students' satisfaction in using wireless internet in higher education: Cross-validation of TSM. *The Electronic Library*, *36*(1), 2–20. https://doi.org/10.1108/EL-07-2016-0150

Jeffrey, L., Hegarty, B., Kelly, O., Penman, M., Coburn, D., & McDonald, J. (2011). Developing digital information literacy in higher education: Obstacles and supports. *Journal of Information Technology Education: Research*, *10*, 383–413.

Jiang, H., Islam, A. Y. M. A., Gu, X., & Spector, J. M. (2021). Online learning satisfaction in higher education during the COVID-19 pandemic: A regional comparison between eastern and western Chinese universities. *Education and Information Technologies*, *26*(6), 6747–6769. https://doi.org/10.1007/s10639-021-10519-x

Jiang, H., Islam, A. Y. M. A., Gu, X., & Spector, J. M. (2022). Technology-enabled e-learning platforms in Chinese higher education during the pandemic age of COVID-19. *SAGE Open*, *12*(2), 1–15. https://doi.org/10.1177/21582440221095085

Joo, S., & Choi, N. (2016). Understanding users' continuance intention to use online library resources based on an extended expectation-confirmation model. *The Electronic Library*, *34*(4), 554–571.

Kandasamy, S., & Vinitha, K. (2014). Online database usage by research scholars of the Manonmaniam Sundaranar University Tirunelveli: A Study. *International Research: Journal of Library and Information Science*, *4*(3), 104–107.

Katabalwa, A. S. (2016). Use of electronic journal resources by postgraduate students at the University of Dar es Salaam. *Library Review*, *65*(6/7), 445–460.

Khan, A. M., Zaidi, S. M., & Zaffar Bharati, M. S. (2009). Use of on-line databases by faculty members and research scholars of Jawaharlal Nehru University (JNU) and Jamia

Millia Islamia (JMI), New Delhi (India): A survey. *The International Information & Library Review*, *41*(2), 71–78.

Koprivica, J., & Grabovac, J. (2010). Search of online data base and information retrieval: One year experience in work of the library of Clinical Centre University in Sarajevo. *Acta Informatica Medica*, *18*(2), 100–108.

Li, H. (2007). Information literacy and librarian-faculty collaboration: A model for success. *Chinese Librarianship*, *24*, 1–14.

Li, X., Islam, A. Y. M. A., Cheng, E. W. L., Hu, X., & Chu, S. K. W. (2022). Exploring determinants influencing information literacy with activity theory. *Online Information Review*, *46*(3), 568–589. https://doi.org/10.1108/OIR-03-2020-0092

Li, Z., Islam, A. Y. M. A., & Spector, J. M. (2022). Unpacking mobile learning in higher vocational education during the COVID-19 pandemic. *International Journal of Mobile Communications*, *20*(2), 129–149. https://doi.org/10.1504/ijmc.2023.10042533

Liu, C. H., & Huang, Y. M. (2015). An empirical investigation of computer simulation technology acceptance to explore the factors that affect user intention. *Universal Access in the Information Society*, *14*(3), 449–457.

Machimbidza, T., & Mutula, S. (2020). Investigating disciplinary differences in the use of electronic journals by academics in Zimbabwean state universities. *Journal of Academic Librarianship*, *46*(2), 1–10.

Masrek, M. N., & Gaskin, J. E. (2016). Assessing users satisfaction with web digital library: The case of Universiti Teknologi MARA. *The International Journal of Information and Learning Technology*, *33*(1), 36–56.

Mehta, D., & Wang, X. (2020). COVID-19 and digital library services – a case study of a university's library. *Digital Library Perspectives*, *36*(4), 351–363. https://doi.org/10.1108/DLP-05-2020-0030

Mohammed, F., Ibrahim, O., Nilashi, M., & Alzurqa, E. (2017). Cloud computing adoption model for e-government implementation. *Information Development*, *33*(3), 303–323.

Pakistan HEC National Digital Library. (2017). *HEC national digital library programme*. www.digitallibrary.edu.pk/Index.php

Peritz, B. C., Wolman, Y., & Bar-Ilan, J. (2003). A survey on the use of electronic databases. *Journal of Academic Librarianship*, *29*(6), 346–361.

Poong, Y. S., Yamaguchi, S., & Takada, J. I. (2017). Investigating the drivers of mobile learning acceptance among young adults in the World Heritage town of Luang Prabang, Laos. *Information Development*, *33*(1), 57–71.

Rafi, M., Ahmad, K., Naeem, S. B., Khan, A. U., & JianMing, Z. (2020). Knowledge-based society and emerging disciplines: A correlation of academic performance. *The Bottom Line*, *33*(4), 337–358. https://doi.org/10.1108/BL-12-2019-0130

Rafi, M., Islam, A. Y. M. A., Ahmad, K., & JianMing, Z. (2022). Digital resources integration and performance evaluation under the knowledge management model in the academic libraries. *Libri – International Journal of Libraries and Information Studies*, *72*(2), 123–140. https://doi.org/10.1515/libri-2021-0056

Rafi, M., JianMing, Z., & Ahmad, K. (2019). Evaluating the impact of digital library database resources on the productivity of academic research. *Information Discovery and Delivery*, *47*(1), 42–52. https://doi.org/10.1108/IDD-07-2018-0025

Rasch, G. (1960). *Probabilistic models for some intelligence and attainment tests*. Copenhagen: Nielsen & Lydiche.

Russell, P. (2008). Information literacy support for off-campus students by academic libraries in the Republic of Ireland. *Journal of information literacy*, *2*(2), 46–62.

Said, A. (2006). *Accessing electronic information: A study of Pakistan's digital library*. INSAP.

Shittu, A. T., Gambari, A. I., Gimba, W. R., & Ahmed, H. (2016). Modeling technology preparedness as an antecedent of mathematic pre-service teachers' self-efficacy, perceived usefulness and intention toward use of information technology in Nigeria. *Malaysian Online Journal of Educational Sciences*, *4*(3), 39–48.

Singh, A., & Gautam, J. N. (2004). Electronic databases: The Indian scenario. *The Electronic Library*, *22*(3), 249–260.

Sinh, N. H., & Nhung, H. T. H. (2012). Users' searching behaviour in using online databases at Vietnam National University – Ho Chi Minh City. *Library Management*, *33*(8), 458–468.

Sinha, A. K. (2016). *Digital information literacy of post graduate students of Visva-Bharati: A survey*. http://ir.inflibnet.ac.in/handle/1944/2022

Sobel, M. E. (1982). Asymptotic confidence intervals for indirect effects in structural equation models. *Sociological Methodology*, *13*, 290–312.

Stewart, R., Narendra, V., & Schmetzke, A. (2005). Accessibility and usability of online library databases. *Library Hi Tech*, *23*(2), 265–286.

Suki, N. M. (2016). Willingness of patrons to use library public computing facilities: Insights from Malaysia. *The Electronic Library*, *34*(5), 823–845.

Teo, T., & Zhou, M. (2014). Explaining the intention to use technology among university students: A structural equation modeling approach. *Journal of Computing in Higher Education*. *26*(2), 124–142.

Tlakula, T. P., & Fombad, M. (2017). The use of electronic resources by undergraduate students at the University of Venda, South Africa. *The Electronic Library*, *35*(5), 861–881.

Tripathi, M., & Kumar, S. (2014). Use of online resources at Jawaharlal Nehru University: A quantitative study. *Program*, *48*(3), 272–292.

Ukachi, N. B. (2015). Information literacy of students as a correlate of their use of electronic resources in university libraries in Nigeria. *The Electronic Library*, *33*(3), 486–501.

Verma, S. (2016). Use of online databases in central science library, University of Delhi: A survey. *DESIDOC Journal of Library & Information Technology*, *36*(2), 104–107.

Yan, W., Deng, S., & Zhang, Y. (2016). Factors influencing the intention to use information service mashups: An empirical study of digital libraries in China. *The Electronic Library*, *34*(4), 696–716.

Yuan, S., Liu, Y., Yao, R., & Liu, J. (2016). An investigation of users' continuance intention towards mobile banking in China. *Information Development*, *32*(1), 20–34.

Appendix

This questionnaire attempts to investigate "**Assessing Online Research Databases in Higher Education Using TSM**" "**Online research databases provided by the COMSATS University Islamabad (CUI) library accessible from its website. The databases comprise huge volumes of research articles, e-books, theses and reports published in journals as well as conference proceedings**".

Direction

The questionnaire has five sections. For all multiple-choice questions, please indicate your response by placing a tick (/) in the appropriate box. For five-point Likert scale questions, please circle the number of your choice. If you wish to comment on any question or qualify your answer, please feel free to use the space in the margin or write your comments on a separate sheet of paper. All information that you provide will be kept strictly confidential. Thank you!

Section I: Demographic information

1. Gender: Male   ☐
               Female   ☐

2. Age: 21–25 years old   ☐
             26–30 years old   ☐
             31–35 years old   ☐
             36 years old and above ☐

3. Number of years studying at the CUI
   a. 0–1 years
   b. 2–3 years
   c. 4–5 years
   d. 6–7 years
   e. 8 years and above

4. Level of study
   a. Master
   b. PhD

102 *Applying the Rasch Model and Structural Equation Modeling to Higher Education*

5. Faculty_____

6. Nationality_____

Section II: Perceived ease of use of online research databases

For questions 1–10, rate how much you agree with each statement using the following scale:

1 = Strongly Disagree (SD)   4 = Agree (A)
2 = Disagree (D)   5 = Strongly Agree (SA)
3 = Neither agree nor disagree (N)

No.	Items	SD	D	N	A	SA
1	I find the university online research databases easy to use.	1	2	3	4	5
2	I find it easy to access the university online research databases.	1	2	3	4	5
3	It is easy for me to become skilful at using online research databases.	1	2	3	4	5
4	It is easy for me to remember how to search journals/articles by using the online research databases.	1	2	3	4	5
5	Interacting with the online research databases system requires minimal mental effort.	1	2	3	4	5
6	I do not feel the need to consult any user manual when using online research databases.	1	2	3	4	5
7	I find it easy to get the online research databases to help me do my research.	1	2	3	4	5
8	My interaction with the online research databases is within my comprehension.	1	2	3	4	5
9	Interacting with the online research databases system is very stimulating for me.	1	2	3	4	5
10	I find it easy to choose articles/journals of different categories, for example, education, engineering and business.	1	2	3	4	5

Section III: Perceived usefulness of online research databases

For questions 1–10, rate how much you agree with each statement using the following scale:

1 = Strongly Disagree (SD)   4 = Agree (A)
2 = Disagree (D)   5 = Strongly Agree (SA)
3 = Neither agree nor disagree (N)

No.	Items	SD	D	N	A	SA
1	Using the online research databases enables me to get research materials.	1	2	3	4	5
2	Using the online research databases helps me write my thesis/journal articles.	1	2	3	4	5
3	The online research database addresses my thesis-related needs.	1	2	3	4	5

No.	Items	SD	D	N	A	SA
4	My job would be difficult to perform without the online research databases.	1	2	3	4	5
5	Using the online research databases system saves me time.	1	2	3	4	5
6	Using the online research databases system allows me to accomplish more work than would otherwise be possible.	1	2	3	4	5
7	Using the online research databases improves the quality of my work.	1	2	3	4	5
8	Using the online research databases enhances my knowledge and research skills.	1	2	3	4	5
9	Using the online research databases increases my research productivity.	1	2	3	4	5
10	Using the online research databases provides me with the latest information on particular areas of research.	1	2	3	4	5

Section IV: Computer self-efficacy of online research databases

For questions 1–8, rate how much you agree with each statement using the following scale:

1 = Strongly Disagree (SD) 4 = Agree (A)
2 = Disagree (D) 5 = Strongly Agree (SA)
3 = Neither agree nor disagree (N)

No.	Items	SD	D	N	A	SA
1	I am able to use the online research databases.	1	2	3	4	5
2	I can download research materials from the online research databases.	1	2	3	4	5
3	I can navigate my way through the online research databases.	1	2	3	4	5
4	I can use the online research databases to enhance the quality of my research.	1	2	3	4	5
5	I can save and print journals/articles from the online research databases.	1	2	3	4	5
6	I have the ability to e-mail journals/articles from online research databases.	1	2	3	4	5
7	I have the knowledge and skills required to benefit from using the online research databases.	1	2	3	4	5
8	I can access the online research databases from the university website.	1	2	3	4	5

Section V: Satisfaction of online research databases

For questions 1–5, rate the likelihood of each using the following scale:

1 = Very Unsatisfied (VU) 4 = Satisfied (S)
2 = Unsatisfied (U) 5 = Very Satisfied (VS)
3 = Not Sure (NS)

No.	Items	SD	D	N	A	SA
1	Overall, I am satisfied with the ease of completing my task by using online research databases.	1	2	3	4	5
2	The university online research databases service has greatly affected the way I search for information and conduct my research.	1	2	3	4	5
3	Providing the online research databases is an indispensable and satisfactory service provided by the university.	1	2	3	4	5
4	Overall, I am satisfied with the amount of time it takes to complete my task by using online research databases.	1	2	3	4	5
5	I am satisfied with the structure of accessible information (available as categories of a research domain, or by the date of issue – for journals in particular – or as full-texts or abstracts of theses and dissertations) of the online databases.	1	2	3	4	5

4 Measurement of Wireless Internet in Higher Education Using the TSM

4.1 Introduction

The late twentieth century was a time of major changes in human history characterised by the shift from the industrial age to the information age wherein knowledge is created, transferred, consumed and re-created at exponential rates. These changes were enabled by rapid expansions in the internet and information and communication technologies (ICTs). The impacts of ICT have been observed in education (Islam et al., 2019; Islam et al., 2020; Islam et al., 2022; Li et al., 2022; Xu et al., 2021; Bin et al., 2020), business, health, industry and almost all other fields. Governments have also been impacted by changes brought about by the digital era, and restructuring of country strategies to manage these changes has become one of the top priorities of governments. This is particularly so in developing countries. In conjunction with this undertaking, developing countries recognised that their long-term target is to achieve an information culture by encouraging ICT access and by ensuring diffusion of technology services to all citizens (Afacan et al., 2013).

In line with technological developments, there have been major changes in the education sector. Notably, wireless internet technology is now widely available as an essential educational resource in various forms of electronic learning (e-learning) and mobile learning (m-learning) in higher education in both developed and developing countries, including Australia, Malaysia, Turkey, the UK and the USA (Alsabawy et al., 2013; Chuang et al., 2015; Chun, 2014; Ilgaz & Gülbahar, 2015; Islam et al., 2015; Shin & Kang, 2015; Yu, 2015; Jiang et al., 2021, 2022; Li et al., 2021; Li et al., 2022). Recent research found that many students are frequent users of internet-based devices to facilitate knowledge acquisition (Chang et al., 2014; Chuang et al., 2015; Kuo et al., 2014). Numerous studies have identified empirical evidence in support of positive impacts on students who learn through wireless internet, including physical mobility (Mohammadi, 2015), enhanced metacognition (Yu & Liu, 2014), opportunities offered for self-monitoring and self-regulation in the learning process (Hamid et al., 2015), enhanced interaction with lecturers (Hamid et al., 2015) and the flexibility of being able to study anywhere at any time, a feature important to adults, particularly women, in countries where, for cultural or

DOI: 10.1201/9781003384724-4

other reasons, they previously did not have access to traditional higher education (Ilgaz & Gülbahar, 2015; Mohammadi, 2015).

Despite the ubiquitous wireless internet applications in higher education, there is a dearth of research into contributing factors of wireless internet user satisfaction within the higher education context (Alsabawy et al., 2013; Ilgaz & Gülbahar, 2015; Kim-Soon et al., 2014; Mohammadi, 2015). Research into student satisfaction and factors impacting levels of satisfaction in using wireless internet-based learning resources is important for course designers and educators in order to target areas for change and improvement, so as to enhance the quality of the learning environment by scientific, valid and reliable approaches (Alsabawy et al., 2013; Ilgaz & Gülbahar, 2015; Kuo et al., 2014). Learner satisfaction is an important outcome and an indicator of online programme implementation (Kuo et al., 2014; Jiang et al., 2021, 2022; Li et al., 2021). Higher student satisfaction was found to be associated with improved student retention, increased academic persistence, higher use and adoption rate (Chen et al., 2020; Jiang et al., 2022) and enhanced academic success in the programme (Kim-Soon et al., 2014; Kuo et al., 2014; Liao & Hsieh, 2011). Hence, the primary aims of this study are to examine factors that influence student satisfaction in using wireless internet in higher education for learning purposes and to identify whether gender has a moderating effect on student satisfaction. To reach the aims of the study, this research paper first reviews the literature on the technology satisfaction model (TSM) and constructs hypotheses. The next section defines the methodology and follows with results from extensive data analyses. The chapter concludes with a discussion, provides a number of practical, methodological and theoretical implications and also sheds light on future studies.

4.2 Background

Many researchers have been extending the technology acceptance model (TAM) by incorporating additional factors that can facilitate its generalisability (Ahmad et al., 2010; Al-Ruz & Khasawneh, 2011; Alenezi et al., 2010; Chitungo & Munongo, 2013; Hanafizadeh et al., 2014; Lee et al., 2011; Rasimah et al., 2011; Shin, 2012; Shittu et al., 2013; Stone and Baker-Eveleth, 2013; Sun & Zhang, 2006; Terzis et al., 2013; Tezci, 2011; Thompson et al., 2006; Wong et al., 2012; Zejno & Islam, 2012). However, the literature indicates that very few studies have been conducted by adding satisfaction into the original TAM (Davis et al., 1989) or diffusion of innovation theory (Devaraj et al., 2002; Kumar & Ravindran, 2012; Lee & Lehto, 2013; Lee & Park, 2008; Rogers, 1995; Shipps & Phillips, 2013).

We applied the TSM as a theoretical framework of this study as shown in Figure 4.1. Here, satisfaction is defined as the extent to which the use of technology is consistent with students' existing values, needs and experiences in using wireless internet. The TSM, as developed and validated by Islam (2014), consists of four dimensions: perceived usefulness (PU), satisfaction (SAT),

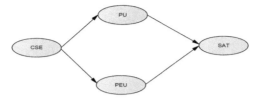

Figure 4.1 Technology satisfaction model (adapted from Islam, 2014).

Note: computer self-efficacy (CSE), satisfaction (SAT), perceived usefulness (PU) and perceived ease of use (PEU).

computer self-efficacy (CSE) and perceived ease of use (PEU). On the other hand, the TAM proposed by Davis et al. (1989) ignored the issue of CSE and SAT (Islam, 2014).

Islam (2014) confirmed that computer self-efficacy is the most dominant antecedent in the TSM while assessing student satisfaction in using wireless internet at the tertiary level of education. Islam's (2014) study also disclosed that the influence of students' computer self-efficacy on satisfaction, perceived ease of use and usefulness were limited to their capabilities and requisite skills in using wireless internet, an online research database, saving and printing journal articles and accessing databases from the university website. However, students outlined the obstacles they undergo while accessing journals from the university using online research databases.

Subsequently, Islam et al. (2015) examined the efficacy of the TSM on postgraduate students' satisfaction when they access online research databases in tertiary educational institutions. Their findings showed that the TSM was feasible for evaluating the fruitful incorporation of research databases among postgraduate students. Islam and Sheikh (2020, p. 1) found, "the TSM shows that 69% of the variance in the postgraduate students' satisfaction can be explained by perceived usefulness, computer self-efficacy and perceived ease of use, which confirm that the TSM has a strong predictive power and is viable for measuring the endogenous variable". In a recent study, Islam et al. (2020, p. 14) claimed, "the TSM has 'strong predictive power' in gauging Chinese lecturers' satisfaction with ICT". They also stated, "satisfaction is a complicated concept to define, and this helps provide a model for doing so". Jiang et al. (2021, p. 18) asserted that their

> study successfully applied the TSM and exhibited the direct and indirect impacts of computer self-efficacy, perceived ease of use and usefulness on university students' satisfaction with online learning platforms. It was found that Chinese university students are highly satisfied and that the TSM can powerfully explain and predict Chinese university students' satisfaction with online learning platforms.

Primarily, the TSM was developed and validated in a reputed public university in Malaysia to enhance the credibility of the findings; thus, it calls for inclusion of other universities around the globe that would formulate a more comprehensive study in the area (Islam et al., 2015). As such, for this study we applied TSM as a theoretical framework to investigate Chinese university students' satisfaction when they use wireless internet for their research and learning purposes.

4.3 Literature Review

This section provides relevant literature pertaining to the factors that influence students' satisfaction while using wireless internet in higher education institutions. It then discusses the TSM and its constructs. Based on the review, six hypotheses are posited to validate the TSM and examine students' satisfaction when using wireless internet provided by the university.

Computer self-efficacy refers to students' belief in their computer ability to use wireless internet for their learning and research purposes. The TSM revealed that postgraduate students' satisfaction with online research databases depends on perceived ease of use and usefulness and relies on the belief in their individual capability to use it. Therefore, computer self-efficacy was acknowledged as a distinct antecedent of perceived ease of use, satisfaction and perceived usefulness, which was an important intrinsic motivation attribute of technology satisfaction (Islam, 2014, 2016; Islam et al., 2015; Islam et al., 2019; Islam et al., 2020; Chen et al., 2020; Islam & Sheikh, 2020; Bin et al., 2020; Jiang et al., 2021; Jiang et al., 2022; Du et al., 2022). The TSM confirmed that students' computer self-efficacy had a direct effect on the perceived usefulness and ease of use of wireless internet (Islam, 2014, 2017) and online research databases (Islam et al., 2015; Chen et al., 2020; Islam & Sheikh, 2020; Du et al., 2022). On the other hand, the TSM exhibited that students' computer self-efficacy also articulated an indirect effect on their satisfaction in using wireless internet for learning purposes (Islam, 2014) and databases for their research purposes in higher education (Islam et al., 2015; Du et al., 2022; Islam & Sheikh, 2020; Chen et al., 2020) mediated by perceived usefulness as well as ease of use. Similarly, the technology adoption and gratification (TAG) model (Islam, 2016) recently reported that lecturers' gratification was directly affected by perceived ease of use and usefulness, and it was also indirectly influenced by computer self-efficacy mediated by the perceived ease of use and usefulness of ICT facilities for teaching and research purposes. Notwithstanding, according to Lee and Lehto (2013), YouTube self-efficacy had a moderate effect on usefulness, while pre-service teachers' self-efficacy was the most significant dimension with the strongest direct influence on their technology assimilation (Al-Ruz & Khasawneh, 2011). Alenezi et al. (2010) showed a significant relationship between university students' computer self-efficacy and their intention to use e-learning. In a similar study, Yuena and Ma (2008) hypothesised that computer self-efficacy would have a positive influence on usefulness, ease of use and intention to accept e-learning technology. However, their findings also suggested that computer self-efficacy had a positive effect on only perceived ease of use rather than usefulness and intention to accept an e-learning system. In another study, Chang and Tung (2008) claimed that students' computer self-efficacy had a direct impact on their intention to use the online learning course websites.

In addition, Ahmad et al. (2010) showed that computer self-efficacy indirectly influenced staff use of technology mediated by perceived usefulness and intention to use. In a related study, Wu et al. (2008) depicted a positive relationship among teachers' computer self-efficacy and their intention to use ICT in teaching. Similarly, Mac Callum and Jeffrey (2013) investigated how ICT skills influence university students' adoption of mobile technology in the tertiary learning environment. An extended TAM was applied as a theoretical framework in their study. The hypothesised model included ICT self-efficacy (basic and advanced ICT skills and advanced mobile skills) into the TAM. The findings of their study confirmed that students' behavioural intention to accept mobile learning was affected by ease of use, basic ICT skills and usefulness. However, basic ICT skills and ease of use had a moderate influence on intention to adopt mobile technology. Additionally, basic ICT skills and advanced mobile skills were moderately significant factors of mobile learning adoption. Along this line, Lee et al. (2014) extended the TAM by including additional factors of internet self-efficacy, learning content, computer self-efficacy, instructor attitude and technology accessibility to investigate student adoption of e-learning. The results of their model confirmed that usefulness and ease of use had positive effects on intention to use e-learning, while usefulness and ease of use were positively influenced by learning content. Additionally, computer self-efficacy found a positive effect on ease of use, but it had a negative nonsignificant impact on usefulness. Subsequently, perceived ease of use and usefulness were moderately influenced by internet self-efficacy. However, instructor attitude towards students showed a negative nonsignificant impact on usefulness. Finally, ease of use was moderately influenced by accessibility of technology while it significantly influenced usefulness. Instructors' suggestions for researchers included that the individual characteristics of computer self-efficacy and internet self-efficacy play crucial roles in influencing a user's belief about ease of use. In the present study, these observations led to the following hypotheses:

H1: Computer self-efficacy (CSE) would have a significant direct effect on perceived ease of use (PEU) of wireless internet in higher education.

H2: Computer self-efficacy (CSE) would have a significant direct effect on perceived usefulness (PU) of wireless internet in higher education.

H3: Computer self-efficacy (CSE) would have a significant indirect effect on satisfaction (SAT) mediated by perceived ease of use (PEU) of wireless internet in higher education.

110   *Applying the Rasch Model and Structural Equation Modeling to Higher Education*

H4: Computer self-efficacy (CSE) would have a significant indirect effect on satisfaction (SAT) mediated by perceived usefulness (PU) of wireless internet in higher education.

Perceived ease of use refers to students' perception of how easy it was to use the wireless internet service for their learning and research purposes. The TSM showed that perceived ease of use of wireless internet indicated a direct impact on students' satisfaction (Islam, 2014), as well as using online research databases (Islam et al., 2015). The TAG model (Islam, 2016) also validated the same relationship when utilising ICT facilities. Previously, Islam (2011) revealed that perceived ease of use showed a direct effect on satisfaction in using technology for learning purposes in tertiary education (Islam, 2014, 2016; Chen et al., 2020; Islam & Sheikh, 2020; Bin et al., 2020; Islam et al., 2015; Islam et al., 2019; Islam et al., 2020; Jiang et al., 2021; Jiang et al., 2022; Du et al., 2022). On the other hand, Kumar and Ravindran (2012) hypothesised that there would be a positive relationship between satisfaction and ease of use of mobile banking. However, their findings did not confirm the relationship among these factors. Lee and Lehto (2013) revealed that YouTube users' ease of use positively affects intention to use and usefulness. Nevertheless, usefulness and behavioural intention were not significantly influenced by the ease of use of YouTube for procedural learning, while the importance of ease of use in assuming technology acceptance was significantly highlighted in the TAM (Davis et al., 1989). Several previous studies also concluded that ease of use had a statistically significant impact on user satisfaction in using mobile technology (Lee & Park, 2008), self-service internet technologies (Huang, 2008; Shamdasani et al., 2008) and electronic commerce channels (Devaraj et al., 2002). These observations led to the next hypothesis:

H5: Perceived ease of use (PEU) would have a significant direct effect on satisfaction (SAT) of wireless internet in higher education.

Perceived usefulness refers to students' perception of the benefits derived from using wireless internet. The TSM confirmed that perceived usefulness was considered to be one of the vital constructs, having a significant direct effect on students' satisfaction in using wireless internet technology (Islam, 2014, 2017), online learning (Jiang et al., 2021, 2022) and online research databases for their learning and research purposes in higher education institutions (Islam et al., 2015; Du et al., 2022; Chen et al., 2020; Islam & Sheikh, 2020). Islam (2016) introduced the TAG model in examining Malaysian and Chinese lecturers' ICT use for their teaching and research purposes where the usefulness of ICT facilities confirmed a direct effect on gratification. According to Islam (2011), the effect of perceived usefulness on satisfaction exhibited the low path coefficient of 0.19, even though its effect on satisfaction could be reflected by the interrelationships among other dimensions, such as computer self-efficacy and ease of use of wireless internet. Kumar and Ravindran (2012) and Lee and Park (2008) explicated that usefulness showed a significant effect on satisfaction. According to Shipps and Phillips (2013), perceived usefulness had an impact on their satisfaction in browsing social networking websites, whereas $\beta = 0.16$ indicates a low path coefficient. In other studies, researchers indicated that perceived usefulness was confirmed as having a significant direct effect on user satisfaction in using YouTube (Lee & Lehto, 2013) and electronic commerce channels (Devaraj et al., 2002). However, Huang (2008) revealed that perceived usefulness had an influence on consumers' satisfaction mediated by their behavioural attitudes. This led to the next hypothesis that reads:

> *H6*: Perceived usefulness (PU) would have a significant direct effect on satisfaction (SAT) with wireless internet.

4.4 Methodology

After the instrument draft was completely translated, a pilot test was conducted with 40 items. In total, 100 students from five colleges of Jiaxing University participated, and the Rasch model was used for analysis. There was consistency between the results of the pilot test and the findings of the ultimate reliability and validity test of the questionnaire. From November to December 2014, 300 students equally distributed among five colleges (Foreign Language Studies, Business, Education, Biology and Chemistry, and Mathematics and Engineering) of Jiaxing University were sampled using the quota sampling technique. The inclusion criterion used was students with personal laptops and a wireless internet connection. However, 17 participants were removed from the original pool of data because of incomplete responses. Thus, in total, 283 students were involved in the final study. All participants were undergraduate students, of whom 70% were female and 30% were male. Most of them (90%) were between 21 and 30 years old. The sample size of this study was deemed sufficient for applying structural equation modeling (SEM) to validate the TSM as supported by Hair et al. (2010). Data were analysed using SPSS version 21.0. The TSM was estimated using SEM through AMOS software version 18.0. The reliability and validity of the measurement scale were tested through a Rasch model using Winsteps version 3.94.

4.4.1 Instrument

A set of questionnaires containing 40 validated items from a prior study (Islam, 2014) was adapted and translated to suit the current study. The survey instrument was translated from English to Chinese and Chinese to English; the accuracy of translation was evaluated by experts who were professional translators from the College of Foreign Language Studies at Jiaxing University. This study used a two-part questionnaire as the primary instrument to collect data. The first part consisted of 36 items intended to measure the constructs (computer self-efficacy, perceived ease of use, and perceived usefulness) on a five-point

112   *Applying the Rasch Model and Structural Equation Modeling to Higher Education*

Likert-type agreement scale (1 for "strongly disagree" and 5 for "strongly agree"). The first ten questions required the participants to rate the wireless internet's ease of use. The next ten questions addressed the participants' perceptions of wireless internet usefulness. The subsequent ten questions addressed the respondents' computer self-efficacy. To assess the endogenous variable (satisfaction to utilise wireless internet), the study used a five-point Likert scale survey questionnaire (1 being "very unsatisfied" and 5 being "very satisfied"). The last six questions concerned the participants' satisfaction with wireless internet. The second part of the instrument consisted of four items regarding the participants' demographic information, which included their gender, age, level of study and college/faculty.

4.4.1.1 Instrument Reliability and Validity

In establishing the validity and reliability of the questionnaire items, a Rasch model was used. According to Liu and Boone (2006), the Rasch model has been progressively applied in a broad range of disciplines for more than three decades. It is a theory-based method for developing assessments. Developing an assessment begins with a clear explanation of the factors or content components to be estimated. The initially predicted determinant or content construct is examined by fitting evaluation data to a Rasch measurement. If data fit the model, then there is substantiation to assert that the initially assumed construct exists, and it is measured by the psychometric properties, hence proffering evidence for construct and content validity. The Rasch model also allows for numerous ways of testing differential item functioning or item bias, which improves the likelihood of developing a fair measurement scale. Applying Winsteps version 3.49, an initial analysis was estimated with 36 items. The findings of the Rasch analyses revealed in the summary statistics were that the items and persons reliability indexes were indicated to be .98 and .89, respectively, whereas the persons and items separation were 2.85 and 7.75, respectively. These results of summary statistics were statistically significant (see Figure 4.2).

Figure 4.3 exhibits the results on an item map where all the items are calibrated on a single continuum. As indicated on the right-hand side of Figure 4.3, the indicators on this scale are arranged from the "less difficult to be rated as students' satisfaction with wireless internet" to the "more difficult to be rated as students' satisfaction with wireless internet". The left-hand side of Figure 4.3 also aligns with the level of a person's ability on a single scale, whereby the students are ordered from the "higher level of ability to endorse students' satisfaction with wireless internet" to the "lower level of ability to endorse students' satisfaction with wireless internet". The findings from the item map also exhibit that the most difficult item was PU4, while the least difficult item was SE8, as reported in Figure 4.3.

The item polarity map exhibited all the items estimated in a similar direction as shown by the point measure correlation (PTMEA CORR.) of > 0.34; however, the only exception was a single item (PEU6), which had a PTMEA

```
Figure          Wireless Internet                      ZOU004ws.txt Apr 21 14:53
INPUT: 283 PERSONS, 36 ITEMS MEASURED: 283 PERSONS, 36 ITEMS, 5 CATS        3.49
-------------------------------------------------------------------------------

    SUMMARY OF 283 MEASURED PERSONS
+-----------------------------------------------------------------------------+
|          RAW                            MODEL      INFIT        OUTFIT       |
|          SCORE    COUNT    MEASURE      ERROR    MNSQ   ZSTD   MNSQ   ZSTD   |
|-----------------------------------------------------------------------------|
| MEAN     121.1    36.0        .33        .19     1.01   -.3    1.00   -.3    |
| S.D.      17.3     .0         .65        .02      .56   2.3     .57   2.3    |
| MAX.     162.0    36.0       2.33        .29     3.61   7.5    3.80   7.9    |
| MIN.      50.0    36.0      -2.39        .18      .18  -5.4     .18  -5.4    |
|-----------------------------------------------------------------------------|
| REAL RMSE    .22 ADJ.SD    .61 SEPARATION  2.85  PERSON RELIABILITY   .89    |
|MODEL RMSE    .19 ADJ.SD    .62 SEPARATION  3.19  PERSON RELIABILITY   .91    |
| S.E. OF PERSON MEAN = .04                                                    |
+-----------------------------------------------------------------------------+
PERSON RAW SCORE-TO-MEASURE CORRELATION = .99
CRONBACH ALPHA (KR-20) PERSON RAW SCORE RELIABILITY = .91

    SUMMARY OF 36 MEASURED ITEMS
+-----------------------------------------------------------------------------+
|          RAW                            MODEL      INFIT        OUTFIT       |
|          SCORE    COUNT    MEASURE      ERROR    MNSQ   ZSTD   MNSQ   ZSTD   |
|-----------------------------------------------------------------------------|
| MEAN     952.0    283.0       .00        .07     1.00   -.2    1.00   -.1    |
| S.D.     124.2     .0         .56        .01      .24   2.9     .25   3.1    |
| MAX.    1141.0    283.0      1.07        .08     1.56   6.5    1.64   6.9    |
| MIN.     698.0    283.0      -.95        .06      .63  -5.2     .60  -5.5    |
|-----------------------------------------------------------------------------|
| REAL RMSE    .07 ADJ.SD    .56 SEPARATION  7.75  ITEM    RELIABILITY  .98    |
|MODEL RMSE    .07 ADJ.SD    .56 SEPARATION  8.11  ITEM    RELIABILITY  .99    |
| S.E. OF ITEM MEAN = .10                                                      |
+-----------------------------------------------------------------------------+
```

Figure 4.2 The summary statistics.

CORR. of 0.22. Item fit order indicated that the majority of items showed adequate item fit and were constructed on a continuum of increasing intensity. The only exception was item PEU6, which is slightly larger than the suggested value. According to Bond and Fox (2001), infit and outfit mean square (MNSQ) statistics for the items are recommended to be larger than 0.5 and smaller than 1.5 for rating scale application, while MNSQ statistics for Pearson measures are suggested to be <1.8. As a result, the misfitting item PEU6 was removed; thus, a total of 35 items were used for the SEM analysis (see Table 4.1).

4.5 Results

SEM is one of the statistical models that seeks to validate, extend and propose new theories. In other words, SEM is also a versatile analytical tool to analyse the mediating and moderation variables and effects. The mediating variable will mediate the relationship between exogenous and endogenous variables. For example, we tested how PU and ease of use mediate the relationship between

114 *Applying the Rasch Model and Structural Equation Modeling to Higher Education*

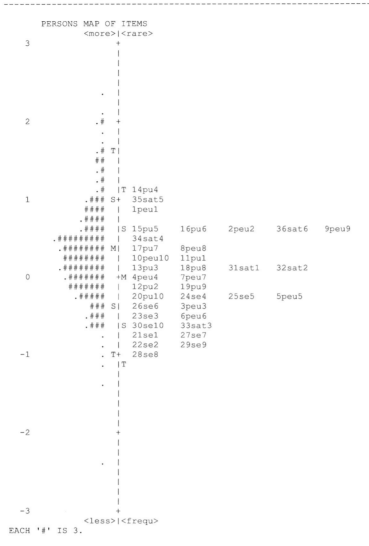

Figure 4.3 Item map.

CSE and SAT. The moderating variable will moderate the causal relationships among the constructs. For instance, gender as a moderating variable was examined to see how it affects the relationship within the TSM. The three-stage findings of the SEM that address the aims of the study are illustrated in this section. First, confirmatory factor analysis (CFA) was applied to validate the

Measurement of Wireless Internet in HE Using the TSM   115

Table 4.1 The 35 Valid Items

Constructs	Valid Items		α
PEU	PEU1	I find it comfortable to use wireless internet in terms of speed.	.702
	PEU2	I find it easy to connect to wireless internet.	
	PEU3	I find it easy to access learning materials from internet.	
	PEU4	I find it easy to access online databases to do research.	
	PEU5	It is easy for me to become skillful in surfing the internet using wireless facility.	
	PEU7	I find it easy to easy to register to use wireless internet.	
	PEU8	I find it easy to complete the online course registration using wireless internet.	
	PEU9	Interacting with wireless internet is very stimulating for me.	
	PEU10	I find it easy to use wireless internet for learning purposes.	
PU	PU1	Using wireless internet improves my learning.	.838
	PU2	Using wireless internet enables me to download learning materials from the internet.	
	PU3	Wireless internet allows me to access databases to do what I want to do.	
	PU4	My study would be difficult to perform without wireless internet.	
	PU5	Using wireless internet saves time.	
	PU6	Using the wireless internet system allows me to accomplish more work than would otherwise be possible.	
	PU7	Wireless internet helps me access online databases to enhance my research.	
	PU8	Using wireless internet allows me to obtain multimedia facilities.	
	PU9	Wireless internet enables me to browse different websites.	
	PU10	Using wireless internet allows me to e-mail and chat with others.	
CSE	CSE1	I feel capable of using wireless internet.	.883
	CSE2	I have the ability to download learning/research materials from the internet.	
	CSE3	I believe I have the ability to navigate my way through wireless internet.	
	CSE4	I have the skills required to use wireless internet to enhance the effectiveness of my learning.	
	CSE5	I have the ability to save and print journals/articles from online databases using wireless internet.	
	CSE6	I can easily go through the steps of downloading software.	
	CSE7	I am competent in going through the steps of online buying.	
	CSE8	I am capable of reading news online.	
	CSE9	I have the ability to fill out online surveys/questionnaires.	
	CSE10	I have the ability to do blogging using wireless internet.	
SAT	SAT1	Overall, I am satisfied with the ease of completing my tasks using wireless internet.	.755
	SAT2	The wireless internet service has greatly affected the way I search for information and manage my studies.	
	SAT3	Wireless internet is an indispensable service provided by the university.	
	SAT4	Overall, I am satisfied with the wireless service provided at the university.	
	SAT5	I am satisfied with using wireless internet anytime and anywhere on campus.	
	SAT6	I am satisfied with accessing wireless internet from the dormitory.	

116 *Applying the Rasch Model and Structural Equation Modeling to Higher Education*

Table 4.2 Fit Statistics of the Measurement Models

Measurement Models	χ^2	df	p	RMSEA	CFI	TLI
CFA-1 (SAT)	2.271	1	0.132	0.067	0.996	0.976
CFA-2 (PEU)	19.514	13	0.108	0.042	0.975	0.960
CFA-3 (CSE)	53.532	20	0.000	0.077	0.958	0.941
CFA-4 (PU)	42.241	14	0.000	0.085	0.944	0.916

four measurement models of the TSM. Second, the TSM was estimated using the full-fledged structural model. Third, cross-validation of the TSM was tested using invariance analyses (configural and metric invariance).

The present study validated four measurement models of the TSM, namely, SAT, PEU, CSE and PU applying CFA. In each case, in the modified four CFA models, a few indicators were deleted one at a time which produced manifold large residuals and associated with the biggest modification indices with comparatively low factor loadings, and that were exhibited to be cross-loaded. The overall fit statistics of the modified four CFA models confirmed a satisfactory fit to the data as depicted in Table 4.2. The findings of these measurement models fit indices were recommended by prior studies (Hu & Bentler, 1999; Marsh et al., 2004).

4.5.1 Estimating the Technology Satisfaction Model

The TSM consisted of four measurement models, namely, CSE, PEU, PU and SAT, and was estimated using SEM. According to Byrne (2000), the estimation of parameters can be applied using the maximum likelihood technique provided by AMOS 18 to evaluate the model fit statistics, measurement and structural models. The findings of the TAG model confirmed that all the path coefficients (β) were statistically significant, thus validating the hypotheses. The items of the TSM contained neither any negative loadings nor error variances, as the model confirmed that they were free from offending estimations. The model's overall goodness-of-fit statistics showed a satisfactory fit; the model χ^2 ($df = 295$) = 826.981, $p = .000$ and the root mean square error of approximation (RMSEA) showed the model's adequacy at the acceptable value of .080 as shown in Figure 4.4. However, the value of the comparative fit index (CFI) and the Tucker–Lewis index (TLI) were not recommended. Moreover, a few items produced multiple large residuals and, further, they had large modification indices suggesting the presence of multiple collinearities, which thus necessitated the revision of the TSM.

4.5.1.1 The Revised Technology Satisfaction Model

The TSM was re-specified and re-estimated to assess its overall competence. In the revised TSM, we removed a few items, one at a time (PEU1, PEU2,

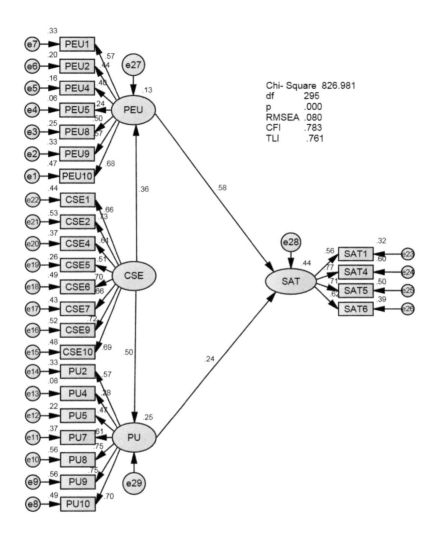

Figure 4.4 Technology satisfaction model.

PEU4, PEU5, CSE2, CSE4, CSE9, PU2, PU4, PU5, PU7, SAT1 and SAT5), which produced multiple large residuals with moderately low factor loadings and that were revealed to be cross-loaded. After dropping these items, only 13 items remained from the original pool of 26 items. The results in Figure 4.3 indicate that the indicators of the revised model were free from negative loadings, error variance and path coefficients. The standardised factor represented loadings ranging from .50 to .77, representing statistically significant indicators. The findings of the TSM's overall goodness-of-fit statistics showed an exceptional fit to the empirical data as indicated by the following fit indices: $\chi^2(df = 62) = 103.808$; $p < .001$; RMSEA = .049; CFI = .954; TLI = .943 as consistent with earlier research (Byrne, 2000).

Figure 4.5 showed the standardised path coefficients of the revised TSM. As estimated, *H1* and *H2* were supported; the results of TSM confirmed that

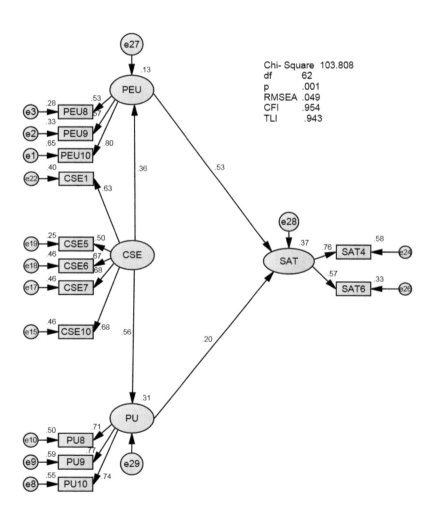

Figure 4.5 The revised technology satisfaction model.

student CSE had a direct effect on the PEU (β = .36, p < 0.001) and PU (β = .56, p < 0.001) of wireless internet in higher education. Consequently, *H3* and *H4* were also supported; the results depicted that student CSE had an indirect effect on their SAT in using wireless internet mediated by PEU (χ^2 = 2.801; p = 0.002), and CSE also indicated an indirect effect on student SAT mediated by their PU of wireless internet (χ^2 = 1.902; p = 0.028) as conducted by the Sobel test (Sobel, 1982). The indirect relationship between CSE and SAT was also practically important. Additionally, *H5* and *H6* were supported, and the findings of TSM exhibited that student PEU (β = .53, p < 0.001) and PU (β = .20, p < 0.015) had a significant direct effect on their SAT in using wireless internet in higher education. In addition, CSE was confirmed to be the most significant antecedent of TSM.

All path coefficients of the revised TSM's critical ratio (CR > 2.439) were statistically significant. Subsequently, the total standardised effect sizes of CSE → PEU and PU were .559 and .356, respectively. Meanwhile, the effect sizes of PEU and PU → SAT were .534 and .202, respectively, while the effect size of CSE → SAT was .303. Finally, the TSM also showed that CSE explained almost 31% and 13% of the variance in PU and PEU, respectively. On the other hand, CSE, PEU and PU together explained approximately 37% of the variability in student SAT in using wireless internet for their learning purposes. Table 4.3 indicates the revised TSM's valid items such as their loading, mean, standard deviation and Cronbach's alpha.

4.5.1.2 Cross-Validation of the Technology Satisfaction Model

The second aim of this study was to identify whether gender has a moderating effect on Chinese student SAT in accessing wireless internet in tertiary education. To assess gender invariance, a two-stage analysis (i.e., configural and metric invariance) was performed on a moderating variable, namely, male (n_1 = 86) and female (n_1 = 197) respondents. First, without constraining the structural paths of the TSM, the findings derived a baseline chi-square value. Afterwards, the structural paths of the model were constrained to be identical for both groups, male and female. The findings for both constrained and unconstrained models were consistent with the data as shown in Figures 4.6 and 4.7.

The metric invariance estimation of this constrained TSM indicated an alternative chi-square (χ^2) value (201.913), which was then evaluated against the unconstrained value (187.782) for statistically significant differences. The findings of invariance analysis across both groups – namely, male and female – resulted in a statistically significant variation in the chi-square value [χ^2 (df = 4) = 14.131, p > .05] as shown in Table 4.4. This meant that gender did exert an effect as a moderating variable towards student SAT in using wireless internet in higher education. It is, thus, reasonable to conclude that there was a significant difference between male and female students in using wireless internet for their learning purposes.

Table 4.3 Valid Items and Their Loadings, Means, Standard Deviations and Alpha Values

Constructs	Items	Item Measure	Loadings	M	SD	Alpha
Computer Self-Efficacy (CSE)	CSE1	I feel capable of using wireless internet.	0.63	3.932	.956	.769
	CSE5	I have the ability to save and print learning materials from the library website using wireless internet.	0.50	3.593	.891	
	CSE6	I can easily go through the steps of downloading software using wireless internet.	0.67	3.685	.962	
	CSE7	I am competent in going through the steps of online buying using wireless internet.	0.68	3.947	.895	
	CSE10	I have the ability to do blogging using wireless internet.	0.68	3.819	.956	
Perceived Usefulness (PU)	PU8	Using wireless internet allows me to obtain multimedia facilities.	0.71	3.311	1.018	.783
	PU9	Wireless internet enables me to browse different websites.	0.77	3.466	1.011	
	PU10	Using wireless internet allows me to e-mail and chat with others.	0.74	3.593	.949	
Perceived Ease of Use (PEU)	PEU8	I find it easy to get the information through the Blackboard (BB) learning system using wireless internet.	0.53	3.074	1.132	.654
	PEU9	Interacting with wireless internet is very stimulating for me.	0.57	2.894	1.032	
	PEU10	I find it easy to use wireless internet for learning purposes.	0.80	3.187	.973	
Satisfaction (SAT)	SAT4	Overall, I am satisfied with the wireless service provided at the university.	0.76	2.929	1.130	.608
	SAT6	I am satisfied with using wireless internet anytime and anywhere on campus.	0.57	2.805	1.294	

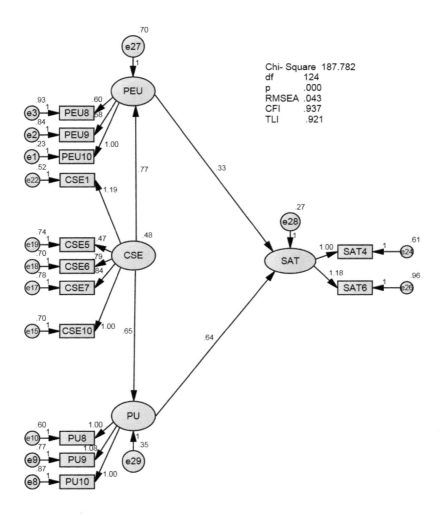

Figure 4.6 The unconstrained model.

4.6 Discussion

The results of the investigation have expanded the existing body of knowledge on student satisfaction with wireless internet and contributed to a better understanding of the TSM in various ways. *H1*, which concludes that students' computer self-efficacy had a significant direct effect on the perceived ease of

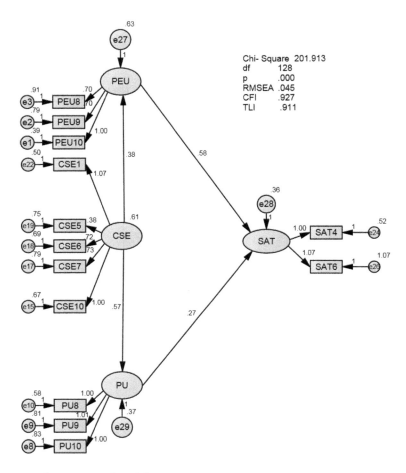

Figure 4.7 The constrained model.

use of wireless internet, stands validated. This hypothesis, however, is coherent with earlier studies which revealed that perceived ease of use was influenced by students' computer self-efficacy in using wireless internet (Islam, 2014) and online research databases (Islam et al., 2015; Chen et al., 2020; Du et al., 2022; Islam & Sheikh, 2020) for learning and research in higher education. This researchers also verified that computer self-efficacy had a substantial effect on perceived ease of use (Islam, 2016, 2017; Islam et al., 2019; Islam et al., 2020; Chen et al., 2020; Islam & Sheikh, 2020; Bin et al., 2020; Jiang et al., 2021, 2022; Du et al., 2022; Lee et al., 2014; Yuena & Ma, 2008). This means that

Table 4.4 Findings of Critical Value of Chi-Squared

Models		Chi-Squared	df	Critical Value	Chi-Squared Change
Gender invariant of TSM	Unconstrained	187.782	124	9.488 ($p > .05$)	14.131
	Constrained	201.913	128 4		

the perceived ease of use of wireless internet to the student depends on their beliefs in their individual capability to use it. This also implies that students' computer self-efficacy related to wireless internet enhances their ease of use. Nevertheless, pre-service teachers' self-efficacy in using technology was the most vital construct with the maximum direct impact on their technology integration (Al-Ruz & Khasawneh, 2011). Correspondingly, Alenezi et al. (2010) claimed a valid relationship between students' computer self-efficacy and their intention to use e-learning in tertiary education institutions.

Second, the results of TSM indicated that students' computer self-efficacy had a significant direct effect on the perceived usefulness of wireless internet in higher education, thereby validating *H2*. This implies that students' computer self-efficacy or ability can lead them to feel the perceived usefulness of wireless internet. For instance, students could gain more benefits based on computer skills by using wireless internet in their academic endeavours. As such, the TSM confirmed that students' computer self-efficacy was the most significant antecedent of the perceived usefulness of using wireless internet for learning in higher education. This was consistent with the existing literature, where several empirical studies (Ahmad et al., 2010; Islam, 2014, 2016; Islam et al., 2015; Lee & Lehto, 2013) also concluded that computer self-efficacy had a direct effect on perceived usefulness. However, Lee et al. (2014) claimed that computer self-efficacy had a negative nonsignificant effect on usefulness, while Yuena and Ma (2008) expressed computer self-efficacy was not a major predictor of usefulness. In a related study, Islam (2011) claimed that when students comprehend the advantages of wireless internet, they are more likely to use the facilities provided by the university, because they believe that wireless internet service can improve their learning performances and outcomes.

Third, the findings of the TSM exhibited statistical support for *H3* and *H4*, which stated that students' computer self-efficacy would exert a positive indirect effect on their satisfaction mediated by the perceived ease of use and usefulness of wireless internet, respectively. To further clarify, this means that students' satisfaction with wireless internet depends on not only perceived usefulness and ease of use but also belief in their personal capability to use it. Although students' self-efficacy does not directly affect their satisfaction in using wireless internet for academic purposes, it could increase their satisfaction through their perception of ease of use and usefulness of the wireless internet service. Thus, computer self-efficacy was recognised as a distinct antecedent of TSM. These results aligned with the results of prior studies that also exhibited students' computer self-efficacy as having a significant indirect effect on their satisfaction in using wireless internet (Islam, 2014) and online research databases for learning and research in higher education (Islam et al., 2015) mediated by the perceived usefulness and ease of use, respectively. The findings of the TAG model proposed by Islam (2016) were also consistent with the TSM but in using lecturers' ICT facilities for their teaching and research purposes. However, Ahmad et al. (2010) evidenced that computer self-efficacy indirectly affected staff's use of computer technology mediated by their usefulness and intention to use in higher education. On the other hand, previous researchers also researched the impact of computer self-efficacy on intention to use (Alenezi et al., 2010; Al-Ruz & Khasawneh, 2011; Mac Callum & Jeffrey, 2013; Wu et al., 2008).

Fourth, the effect of perceived ease of use of wireless internet and students' satisfaction was depicted in a significant direct path coefficient, thereby validating *H5*. This explains that students' satisfaction directly depends on the perceived ease of use of wireless internet service. To enhance students' satisfaction, higher education authorities may need to ensure that the quality of this service remains within the reach of students to facilitate their learning. This was congruent with the findings of prior studies (Devaraj et al., 2002; Huang, 2008; Islam, 2011, 2014, 2016; Islam et al., 2015; Lee & Park, 2008; Shamdasani et al., 2008), which also suggested a significant association between ease of use and satisfaction. This study also affirmed that perceived ease of use of wireless internet was relatively more effective compared to perceived usefulness in affecting students' satisfaction.

Fifth, as regards the influence of perceived usefulness on satisfaction, there was a statistically significant relationship, thereby validating *H6*. This implies that perceived usefulness of wireless internet to the students could increase their satisfaction. In other words, the benefits of the wireless internet service had a positive impact on student perception towards satisfaction of using technology in higher education. This finding corroborates the results of existing empirical studies (Devaraj et al., 2002; Islam, 2014, 2016; Islam et al., 2015; Kumar & Ravindran, 2012; Lee & Lehto, 2013; Lee & Park, 2008; Shipps & Phillips, 2013), which also stated that usefulness had a direct influence on satisfaction. However, Huang (2008) showed that usefulness had an impact on consumer satisfaction through their attitudes.

Finally, the findings of invariance analyses inferred that the TSM elucidates the use of wireless internet among male and female students, as gender seems to constrain the generalizability of the TSM. This clearly implies that, in a setting where university learners have access to wireless internet to enhance their quality of learning, gender diversity affects learners' satisfaction in accessing the wireless internet service. In other words, there was a significant difference between male and female students in using wireless internet in higher education. However, the findings were inconsistent with the findings of previous studies which revealed no significant gender differences with respect to satisfaction (Islam, 2011) and technology adoption (Ahmad et al., 2010; Sam et al., 2005).

4.7 Conclusion

This study has reaffirmed and validated the TSM in another setting and disseminated the SEM and the Rasch model as a method for testing and validating models as well as confirmed the generalizability of the TSM in assessing students' satisfaction with using wireless internet in tertiary education institutions. The psychometric properties of TSM reliability and validity were tested by the Rasch model. The advantage of using the Rasch model was that it measured not only item reliability and validity but also persons, while most studies focused on only item reliability and validity. The valid instrument was used to estimate the TSM by applying a three-stage SEM. The findings of the TSM showed that students' computer self-efficacy indicated not only a direct impact on perceived usefulness and ease of use of wireless internet but also an indirect effect on satisfaction mediated by perceived ease of use and usefulness. Perceived usefulness and ease of use also had a direct effect on students' satisfaction in using the wireless internet service in higher education. The results also confirmed that gender exerted a strong influence as a moderating variable towards students' satisfaction in using wireless internet. These findings proved the generalizability of the TSM. In addition, the revised TSM also showed that computer self-efficacy explained approximately 31% and 13% of the variance in perceived usefulness and ease of use, respectively. Computer self-efficacy, perceived ease of use and usefulness together explained almost 37% of the variability in students' satisfaction in using wireless internet for their academic endeavours. It can be confidently argued that the research findings have some important implications for wireless internet service providers in universities. The study revealed that wireless internet was not convenient to use, especially in terms of perceived ease of use, which accounted for only 13% of the total variance. Service providers would do well to address these issues to better cater to students' needs, thereby enhancing their satisfaction with using wireless internet.

As evidenced from the revised TSM, the direct and indirect effect of students' computer self-efficacy on perceived ease of use, perceived usefulness and satisfaction in using wireless internet for their learning was limited to their capabilities and requisite skills in using wireless internet, saving and printing learning materials from the library website, downloading software, online shopping and blogging. Regarding the effect of perceived usefulness on satisfaction in using wireless internet service, it was manifested in students using multimedia facilities, browsing different websites and communicating with others through e-mail and chat rooms. Concerning the influence of perceived ease of use on learners' satisfaction, this was displayed in terms of their easy access to the information through the Blackboard (BB) learning system using wireless internet, stimulating interaction and easy-to-use wireless internet for learning.

In this study, it was revealed that the TSM is a valid model for examining determinants that affect students' satisfaction in using wireless internet in higher education institutions. The study also identified whether gender has a moderating influence on students' satisfaction while using wireless internet in a developing country, whereas the TSM (Islam, 2014) was developed and validated in the Southeast Asian context. Thus, this research has provided useful information on Jiaxing University students' satisfaction in using the university's wireless internet. These findings can enable the selected university to better gauge the utility of its wireless internet service. Moreover, the practitioners, academicians and researchers may benefit from the findings of the present study; they can apply the TSM in the diverse context of education. Additionally, the extensive analyses and findings also bridge the gap between theory and practice in technological learning and research settings in higher education.

This study had some limitations. It was confined to undergraduate students enrolled in five colleges of a comprehensive public university for foreign language studies, business, education, biology and chemistry, and mathematics and engineering, while many other universities have been implementing the wireless internet service in China. Moreover, this study did not include postgraduate students. Hence, lecturers, administrative staff and postgraduate students might be included in future studies. To generalise the findings, future exploration on cross-cultural validation of the TSM should be performed to shed more light on students' satisfaction in using wireless internet, online research databases, mobile learning, e-learning, information systems, e-portfolios, learning management systems, ICT, social networks, digital library resources and distance education.

4.8 Acknowledgements

This work was supported by the Peak Discipline Construction Project of Education at East China Normal University and Fundamental Research Funds for the Central Universities (2020ECNU-HLYT035).

4.9 References

Afacan, G., Er, E., & Arifoglu, A. (2013). Public internet access points (PIAPs) and their social impact: A case study from Turkey. *Behaviour & Information Technology*, *32*(1), 14–23. https://doi.org/10.1080/0144929X.2011.582149

Ahmad, T. B. T., Basha, K. M., Marzuki, A. M., Hisham, N. A., & Sahari, M. (2010). Faculty's acceptance of computer-based technology: Cross-validation of an extended model. *Australasian Journal of Educational Technology*, *26*(2), 268–279.

Alenezi, A. R., Karim, A. M. A., & Veloo, A. (2010). An empirical investigation into the role of enjoyment, computer anxiety, computer self-efficacy and internet experience in influencing the students' intention to use e-learning: A case study from Saudi Arabian governmental universities. *The Turkish Online Journal of Educational Technology*, *9*(4), 22–34.

Al-Ruz, J. A., & Khasawneh, S. (2011). Jordanian pre-service teachers' and technology integration: A human resource development approach. *Educational Technology & Society*, *14*(4), 77–87.

Alsabawy, A. Y., Cater-Steel, A., & Soar, J. (2013). E-learning service delivery quality: A determinant of user satisfaction. In Y. Kats (Ed.), *Learning management systems and instructional design: Best practices in online education* (pp. 89–127). Information Science Reference. https://doi.org/10.4018/978-1-4666-3930-0.ch006

Bin, E., Islam, A. Y. M. A., Gu, X., Spector, J. M., & Wang, F. (2020). A study of Chinese technical and vocational college teachers' adoption and gratification in new technologies. *British Journal of Educational Technology, 51*(6), 2359–2375. https://doi.org/10.1111/bjet.12915

Bond, T. G., & Fox, C. M. (2001). *Applying the Rasch model: Fundamental measurement in the human science*. Lawrence Erlbaum.

Byrne, B. M. (2000). *Structural equation modeling with Amos: Basic concepts, applications and programming*. Erlbaum.

Chang, C. S., Liu, E. Z. F., Sung, H. Y., Lin, C. H., Chen, N. S., & Cheng, S. S. (2014). Effects of online college student's Internet self-efficacy on learning motivation and performance. *Innovations in Education and Teaching International, 51*(4), 366–377. https://doi.org/10.1080/14703297.2013.771429

Chang, S., & Tung, F. C. (2008). An empirical investigation of students' behavioural intentions to use the online learning course websites. *British Journal of Educational Technology, 39*(1), 71–83. https://doi.org/10.1111/j.1467-8535.2007.00742.x

Chen, H., Islam, A. Y. M. A., Gu, X., Teo, T., & Peng, Z. (2020). Technology-enhanced learning and research using databases in higher education: The application of the ODAS model. *Educational Psychology, 40*(9), 1056–1075. https://doi.org/10.1080/01443410.2019.1614149

Chitungo, S. K., & Munongo, S. (2013). Extending the technology acceptance model to mobile banking adoption in rural Zimbabwe. *Journal of Business Administration and Education, 3*(1), 51–79.

Chuang, S. C., Lin, F. M., & Tsai, C. C. (2015). An exploration of the relationship between internet self-efficacy and sources of internet self-efficacy among Taiwanese university students. *Computers in Human Behavior, 48*, 147–155. http://dx.doi.org/10.1016/j.chb.2015.01.044

Chun, M. (2014). A study on college students' use intention of internet learning resources in Chongqing. *Asian Social Science, 10*(3), 70–78. https://doi.org/10.5539/ass.v10n3p70

Davis, F. Bagozzi, R. P., & Warshaw, P. R. (1989). User acceptance of computer-technology: A comparison of two theoretical models. *Management Science, 38*(8), 982–1003.

Devaraj, S., Fan, M., & Kohli, R. (2002). Antecedents of b2c channel satisfaction and preference: Validating e-commerce metrics. *Information Systems Research, 13*(3), 316–333. http://dx.doi.org/10.1287/isre.13.3.316.77

Du, H., Chen, H., & Islam, A. Y. M. A. (2022). Students' perception of academic databases as recognition of learning and research during the COVID-19 pandemic. *Journal of Information Science*. https://journals.sagepub.com/doi/full/10.1177/01655515221118666

Hair, J. F., Jr., Black, W. C., Babin, B. J., & Anderson, R. E. (2010). *Multivariate data analysis a global perception*. Pearson.

Hamid, S., Waycott, J., Kurnia, S., & Chang, S. (2015). Understanding students' perceptions of the benefits of online social networking use for teaching and learning. *Internet and Higher Education, 26*, 1–9.

Hanafizadeh, P., Behboudi, M., Koshksaray, A. A., & Tabar, M. J. S. (2014). Mobile-banking adoption by Iranian bank clients. *Telematics and Informatics, 31*, 62–78. http://dx.doi.org/10.1016/j.tele.2012.11.001

Hu, L. T., & Bentler, P. M. (1999). Cutoff criteria for fit indexes in covariance structure analysis: Conventional criteria versus new alternatives. *Structural Equation Modeling, 6*, 1–55.

Huang, E. (2008). Use and gratification in e-consumers. *Internet Research, 18*(4), 405–426.

Ilgaz, H., & Gülbahar, Y. (2015). A snapshot of online learners: E-readiness, e-satisfaction and expectations. *International Review of Research in Open and Distributed Learning*, *16*(2), 171–187.

Islam, A. Y. M. A. (2011). Viability of the extended technology acceptance model: An empirical study. *Journal of Information and Communication Technology*, *10*, 85–98.

Islam, A. Y. M. A. (2014). Validation of the technology satisfaction model (TSM) developed in higher education: The application of Structural Equation Modeling. *International Journal of Technology and Human Interaction*, *10*(3), 44–57. https://doi.org/10.4018/ijthi.2014070104

Islam, A. Y. M. A. (2016). Development and validation of the technology adoption and gratification (TAG) model in higher education: A cross-cultural study between Malaysia and China. *International Journal of Technology and Human Interaction*, *12*(3), 78–105. https://doi.org/10.4018/IJTHI.2016070106

Islam, A. Y. M. A. (2017). Technology satisfaction in an academic context: Moderating effect of gender. In A. Mesquita (Ed.), *Research paradigms and contemporary perspectives on human-technology interaction* (pp. 187–211). IGI Global. https://doi.org/10.4018/978-1-5225-1868-6.ch009

Islam, A. Y. M. A., Gu, X., Crook, C., & Spector, J. M. (2020). Assessment of ICT in tertiary education applying structural equation modeling and Rasch model. *SAGE Open*, *10*(4), 1–17. https://doi.org/10.1177/2158244020975409

Islam, A. Y. M. A., Leng, C. H., & Singh, D. (2015). Efficacy of the technology satisfaction model (TSM): An empirical study. *International Journal of Technology and Human Interaction*, *11*(2), 45–60. https://doi.org/10.4018/ijthi.2015040103

Islam, A. Y. M. A., Mok, M. M. C, Gu, X., Spector, J. M., & Leng, C. H. (2019). ICT in higher education: An exploration of practices in Malaysian universities. *IEEE Access*, 7(1), 16892–16908. https://doi.org/10.1109/ACCESS.2019.2895879

Islam, A. Y. M. A., Rafi, M., & Ahmad, K. (2022). Analyzing the impact of technology incentives on the community digital inclusion using structural equation modeling. *Library Hi Tech*. https://doi.org/10.1108/LHT-07-2021-0226

Islam, A. Y. M. A., & Sheikh, A. (2020). A study of the determinants of postgraduate students' satisfaction in using online research databases. *Journal of Information Science*, *46*(2), 273–287. https://doi.org/10.1177/0165551519834714

Jiang, H., Islam, A. Y. M. A., Gu, X., & Spector, J. M. (2021). Online learning satisfaction in higher education during the COVID-19 pandemic: A regional comparison between eastern and western Chinese universities. *Education and Information Technologies*, *26*(6), 6747–6769. https://doi.org/10.1007/s10639-021-10519-x

Jiang, H., Islam, A. Y. M. A., Gu, X., & Spector, J. M. (2022). Technology-enabled e-learning platforms in Chinese higher education during the pandemic age of COVID-19. *SAGE Open*, *12*(2), 1–15. https://doi.org/10.1177/21582440221095085

Kim-Soon, N., Rahman, A., & Ahmed, M. (2014). E-service quality in higher education and frequency of use of the service. *International Education Studies*, 7(3), 1–10. https://doi.org/10.5539/ies.v7n3p1

Kumar, R. G., & Ravindran, S. (2012). An empirical study on service quality perceptions and continuance intention in mobile banking context in India. *Journal of Internet Banking and Commerce*, *17*(1), 1–22.

Kuo, Y. C., Walker, A. E., Schroder, K. E. E., & Belland, B. R. (2014). Interaction, internet self-efficacy, and self-regulated learning as predictors of student satisfaction in online education courses. *Internet and Higher Education*, *20*, 35–50.

Lee, D. Y., & Lehto, M. R. (2013). User acceptance of YouTube for procedural learning: An extension of the technology acceptance model. *Computers & Education*, *61*, 193–208. http://dx.doi.org/10.1016/j.compedu.2012.10.001

Lee, T. M., & Park, C. (2008). Mobile technology usage and b2b market performance under mandatory adoption. *Industrial Marketing Management*, *37*(7), 833–840.

Lee, Y. H., Hsiao, C., & Purnomo, S. H. (2014). An empirical examination of individual and system characteristics on enhancing e-learning acceptance. *Australasian Journal of Educational Technology*, *30*(5), 561–579.

Lee, Y. H., Hsieh, Y. C., & Hsu, C. N. (2011). Adding innovation diffusion theory to the technology acceptance model: Supporting employees' intentions to use e-learning systems. *Educational Technology & Society*, *14*(4), 124–137.

Li, A., Islam, A. Y. M. A., & Gu, X. (2021). Factors engaging college students in online learning: An investigation of learning stickiness. *SAGE Open*, *11*(4), 1–15. https://doi.org/10.1177/21582440211059181

Li, X., Islam, A. Y. M. A., Cheng, E. W. L., Hu, X., & Chu, S. K. W. (2022). Exploring determinants influencing information literacy with activity theory. *Online Information Review*, *46*(3), 568–589. https://doi.org/10.1108/OIR-03-2020-0092

Li, Z., Islam, A. Y. M. A., & Spector, J. M. (2022). Unpacking mobile learning in higher vocational education during the COVID-19 pandemic. *International Journal of Mobile Communications*, *20*(2), 129–149. https://doi.org/10.1504/ijmc.2023.10042533

Liao, P. W., & Hsieh, J. Y. (2011). What influences Internet-based learning? *Social Behavior and Personality*, *39*(7), 887–896.

Liu, X., & Boone, W. (2006). Introduction to Rasch measurement in science education. In X. Liu & W. Boone (Eds.), *Application of Rasch measurement in science education*. JAM Press.

Mac Callum, K., & Jeffrey, L. (2013). The influence of students' ICT skills and their adoption of mobile learning. *Australasian Journal of Educational Technology*, *29*(3), 303–314.

Marsh, H. W., Hau, K. T., & Wen, Z. (2004). In search of golden rules: Comment on hypothesis testing approaches to setting cutoff values for fit indices and dangers in overgeneralizing findings. *Structural Equation Modeling*, *11*, 320–341.

Mohammadi, H. (2015). Investigating users' perspectives on e-learning: An integration of TAM and IS success model. *Computers in Human Behavior*, *45*, 359–374. https://doi.org/10.1016/j.chb.2014.07.044

Rasimah, C. M. Y., Ahmad, A., & Zaman, H. B. (2011). Evaluation of user acceptance of mixed reality technology. *Australasian Journal of Educational Technology*, *27*(8), 1369–1387. www.ascilite.org.au/ajet/ajet27/rasimah.html

Rogers, E. M. (1995). *Diffusion of innovations* (4th ed.). The Free Press.

Sam, H. K., Othman, A. E. A., & Nordin, Z. S. (2005). Computer self-efficacy, computer anxiety, and attitudes toward the internet: A study among undergraduates in Unimas *Educational Technology & Society*, *8*(4), 205–219.

Shamdasani, P., Mukherjee, A., & Malhotra, N. (2008). Antecedents and consequences of service quality in consumer evaluation of self-service internet technology. *The Service Industry Journal*, *28*(1), 117–138.

Shin, D. H. (2012). Factors that influence pre-service teachers' ICT usage in education. *Information Technology & People*, *25*(1), 55–80. https://doi.org/10.1108/09593841211204344

Shin, W. S., & Kang, M. (2015). The use of a mobile learning management system at an online university and its effect on learning satisfaction and achievement. *International Review of Research in Open and Distributed Learning*, *16*(3), 110–130.

Shipps, B., & Phillips, B. (2013). Social networks, interactivity and satisfaction: Assessing socio-technical behavioral factors as an extension to technology acceptance. *Journal of Theoretical and Applied Electronic Commerce Research*, *8*(1), 35–52. https://doi.org/10.4067/S0718–18762013000100004

Shittu, A. T., Basha, K. M., Rahman, N. S. N. A., & Ahmad, T. B. T. (2013). Determinants of social networking software acceptance: A multi-theoretical approach. *The Malaysian Online Journal of Educational Technology, 1*(1), 27–43.

Sobel, M. E. (1982). Asymptotic confidence intervals for indirect effects in structural equation models. *Sociological Methodology, 13*, 290–312.

Stone, R. W., & Baker-Eveleth, L. J. (2013). Students' intentions to purchase electronic textbooks. *Journal of Computing in Higher Education, 25*(1), 27–47. https://doi.org/10.1007/s12528-013-9065-7.

Sun, H., & Zhang, P. (2006). The role of moderating factors in user technology acceptance. *International Journal of Human Computer Studies, 64*(2), 53–78.

Terzis, V., Moridis, C. N., Economides, A. A., & Rebolledo-Mendez, G. (2013). Computer based assessment acceptance: A cross-cultural study in Greece and Mexico. *Educational Technology & Society, 16*(3), 411–424.

Tezci, E. (2011). Factors that influence pre-service teachers' ICT usage in education. *European Journal of Teacher Education, 34*(4), 483–499. https://doi.org/10.1080/02619768.2011.587116

Thompson, R., Compeau, R. D., & Higgins, C. (2006). Intentions to use information technologies: An integrative model. *Journal of Organizational and End User Computing, 18*(3), 25–47.

Wong, K. T., Teo, T., & Russo, S. (2012). Influence of gender and computer teaching efficacy on computer acceptance among Malaysian student teachers: An extended technology acceptance model. *Australasian Journal of Educational Technology, 28*(7), 1190–1207. www.ascilite.org.au/ajet/ajet28/wong-kt.html

Wu, W., Chang, H., & Guo, C. (2008). An empirical assessment of science teacher's intention towards technology integration. *Journal of Computers in Mathematics and Science Teaching, 27*(4), 499–520.

Xu, X., Shen, W., Islam, A. Y. M. A., Shen, J., & Gu, X. (2021). Modeling Chinese teachers' behavioral intention to use recording studios in primary schools. *Interactive Learning Environments.* https://doi.org/10.1080/10494820.2021.1955713

Yu, Z. (2015). Indicators of satisfaction in clickers-aided EFL class. *Frontiers in Psychology, 6*, 1–9. https://doi.org/10.3389/fpsyg.2015.00587

Yu, Z. G., & Liu, C. (2014). The influence of clickers use on metacognition and learning outcomes in College English Classroom. *International Journal of Information and Communication Technology Education, 10*, 50–61. https://doi.org/10.4018/ijicte.2014040105

Yuena, A. H. K., & Ma, W. W. K. (2008). Exploring teacher acceptance of e-learning technology. *Asia-Pacific Journal of Teacher Education, 36*(3), 229–243. https://doi.org/10.1080/13598660802232779

Zejno, B., & Islam, A. (2012). Development and validation of library ICT usage scale for the IIUM postgraduate students. *OIDA International Journal of Sustainable Development, 3*(10), 11–18.

Appendix

The Survey Questionnaire (English version)
This questionnaire attempts to investigate "**Measurement of Wireless Internet in Higher Education Using TSM**"

Direction

The questionnaire has five sections. For all multiple-choice questions, please indicate your response by placing a tick (/) in the appropriate box. For five-point Likert scale questions, please circle the number of your choice. If you wish to comment on any question or qualify your answer, please feel free to use the space in the margin or write your comments on a separate sheet of paper. All information that you provide will be kept strictly confidential. Thank you!

Section I: Perceived ease of use of wireless internet

For questions 1–10, rate how much you agree with each statement using the following scale:

1 = Strongly Disagree (SD)   4 = Agree (A)
2 = Disagree (D)   5 = Strongly Agree (SA)
3 = Neither agree nor disagree (N)

	Statement	*SD*	*D*	*N*	*A*	*SA*
1.	I find it comfortable to use wireless internet in terms of speed.	1	2	3	4	5
2.	I find it easy to connect to wireless internet.	1	2	3	4	5
3.	I find it easy to access learning materials from the internet.	1	2	3	4	5
4.	I find it easy to access online database to do research.	1	2	3	4	5
5.	It is easy for me to become skillful in navigating the internet using wireless facility.	1	2	3	4	5

(Continued)

Statement	SD	D	N	A	SA
6. I do not feel the need to consult any user manual when using wireless internet.	1	2	3	4	5
7. I find it easy to get the registration for using wireless internet.	1	2	3	4	5
8. I find it easy to do the online course registration using wireless internet.	1	2	3	4	5
9. Interacting with the wireless internet is very stimulating for me.	1	2	3	4	5
10. I find it easy to use wireless internet for learning purposes.	1	2	3	4	5

Section II: Perceived usefulness of wireless internet

For questions 11–20, rate how much you agree with each statement using the following scale:

1 = Strongly Disagree (SD) 4 = Agree (A)
2 = Disagree (D) 5 = Strongly Agree (SA)
3 = Neither agree nor disagree (N)

Statement	SD	D	N	A	SA
11. Using the wireless internet improves my learning.	1	2	3	4	5
12. Using the wireless internet enables me to download learning materials from the internet.	1	2	3	4	5
13. Wireless internet allows me to access databases to do what I want to do.	1	2	3	4	5
14. My study would be difficult to perform without wireless internet.	1	2	3	4	5
15. Using wireless internet saves my time.	1	2	3	4	5
16. Using the wireless internet system allows me to accomplish more work than would otherwise be possible.	1	2	3	4	5
17. Wireless internet helps me access online databases to enhance my research.	1	2	3	4	5
18. Using wireless internet allows me to obtain multimedia facilities.	1	2	3	4	5
19. Wireless internet enables me to browse different websites.	1	2	3	4	5
20. Using wireless internet allows me to e-mail and chat with others.	1	2	3	4	5

Measurement of Wireless Internet in HE Using the TSM   133

Section III: Computer self-efficacy of wireless internet

For questions 21–30, rate the likelihood of each using the following scale:

1 = Strongly Disagree (SD)	4 = Agree (A)
2 = Disagree (D)	5 = Strongly Agree (SA)
3 = Neither agree nor disagree (N)	

	Activity	SD	D	N	A	SA
21.	I feel capable of using the wireless internet.	1	2	3	4	5
22.	I have the ability to download learning/research materials from the internet.	1	2	3	4	5
23.	I believe I have the ability to navigate my way through wireless internet.	1	2	3	4	5
24.	I have the skills required to use wireless internet to enhance the effectiveness of my learning.	1	2	3	4	5
25.	I have the ability to save and print journals/articles from online databases using wireless internet.	1	2	3	4	5
26.	I can easily go through the steps of downloading software.	1	2	3	4	5
27.	I am competent in going through the steps of online buying.	1	2	3	4	5
28.	I am capable of reading news online.	1	2	3	4	5
29.	I have the ability to fill out online surveys/questionnaires.	1	2	3	4	5
30.	I have the ability to do blogging using wireless internet.	1	2	3	4	5

Section IV: Satisfaction of wireless internet

For questions 31–35, rate the likelihood of each using the following scale:

1 = Very Unsatisfied (VU)	4 = Satisfied (S)
2 = Unsatisfied (U)	5 = Very Satisfied (VS)
3 = Not Sure (N)	

	Activity	VU	U	N	S	VS
31.	Overall, I am satisfied with the ease of completing my task using wireless internet.	1	2	3	4	5
32.	The wireless internet service has greatly affected the way I search for information and manage my study.	1	2	3	4	5
33.	I am satisfied with wireless service provided by the Information Technology Division (ITD).	1	2	3	4	5

(Continued)

(Continued)

	Activity	VU	U	N	S	VS
34.	Overall, I am satisfied with the wireless service provided at the university.	1	2	3	4	5
35.	I am satisfied in using wireless internet at anytime and anywhere on campus.	1	2	3	4	5
36.	I am satisfied to access Wireless internet from the dormitory.	1	2	3	4	5

Section V: Demographic information

Please complete the following.

37.	Gender:	❑ Male ❑ Female
38.	Age:	_____ yrs
39.	Nationality:	_____
40.	Number of years studying at the JU:	_____ Years.
41.	Level of study:	❑ Undergraduate ❑ Master ❑ PhD
42.	School/Centre:	_____
43.	Department:	_____

Thank you for your gracious cooperation.

Measurement of Wireless Internet in HE Using the TSM　135
(Chinese Version)

本问卷旨在调查"基于技术满意度模型的高等教育中无线网的评价"

指南

问卷有五个部分。对于所有选择题，请在适当的方框内打勾（∕）。对于李克特五分制问题，请圈出你所选的数字。如果你想对任何问题发表评论或更好地进行回答，请随意使用页边空白处的地方，也可以在另一张纸上写下你的评论。您提供的所有资料将被严格保密。谢谢你！

第一部分：对无线网的感知易用性

对于问题1–10，请你用下面的评分对你对每个陈述的同意程度打分：

1 = 非常不同意 (SD)　　　　4 = 同意 (A)
2 = 不同意 (D)　　　　　　5 = 非常同意 (SA)
3 = 一般 (N)

	陈述	SD	D	N	A	SA
1.	就速度而言，我觉得使用无线网很舒服。	1	2	3	4	5
2.	我发现连接无线网络很容易。	1	2	3	4	5
3.	我发现从网上获取学习材料很容易。	1	2	3	4	5
4.	我发现访问在线数据库做研究很容易。	1	2	3	4	5
5.	我很容易掌握使用无线设备浏览互联网的技巧。	1	2	3	4	5
6.	使用无线网时，我觉得不需要查阅用户手册。	1	2	3	4	5
7.	我发现登录使用无线网很容易。	1	2	3	4	5
8.	我发现使用无线网学习在线课程注册很容易。	1	2	3	4	5
9.	我很容易使用无线网。	1	2	3	4	5
10.	我发现使用无线网学习很容易。	1	2	3	4	5

第二部分：对无线网的感知有用性

对于问题11 - 20，请你用下面的评分对你对每个陈述的同意程度打分：

1 = 非常不同意 (SD)　　　　4 = 同意 (A)
2 = 不同意 (D)　　　　　　5 = 非常同意 (SA)
3 = 一般 (N)

陈述	SD	D	N	A	SA
11. 使用无线网提高了我的学习。	1	2	3	4	5
12. 使用无线网可以让我从网上下载学习材料。	1	2	3	4	5
13. 无线网可以让我访问数据库做我想做的事。	1	2	3	4	5
14. 如果没有无线网，我的学习将很难进行。	1	2	3	4	5
15. 使用无线网节省了我的时间。	1	2	3	4	5
16. 使用无线网可以让我完成更多的工作。	1	2	3	4	5
17. 无线网帮助我访问在线数据库，可以改善我的研究。	1	2	3	4	5
18. 使用无线网可以让我获得多媒体设施。	1	2	3	4	5
19. 无线网使我可以浏览不同的网站。	1	2	3	4	5
20. 使用无线网可以让我发电子邮件和和别人聊天。	1	2	3	4	5

第三部分：使用无线网的计算机自我效能

对于问题21 – 30，请你用下面的评分对你对每个陈述的同意程度打分：

1 = 非常不同意 (SD)　　　4 = 同意 (A)
2 = 不同意 (D)　　　　　　5 = 非常同意 (SA)
3 = 一般 (N)

陈述	SD	D	N	A	SA
21. 我觉得我能使用无线网。	1	2	3	4	5
22. 我有能力从网上下载学习/研究资料。	1	2	3	4	5
23. 我相信我有能力使用无线网导航。	1	2	3	4	5
24. 我有能力使用无线网来提高我的学习效率。	1	2	3	4	5
25. 我有能力使用无线网从在线数据库中保存和打印期刊/文章。	1	2	3	4	5
26. 我可以轻松地完成下载软件的步骤。	1	2	3	4	5
27. 我有能力完成网上购物的步骤。	1	2	3	4	5
28. 我能在网上阅读新闻。	1	2	3	4	5
29. 我有能力填写在线调查/问卷。	1	2	3	4	5
30. 我有能力用无线网写博客。	1	2	3	4	5

第四部分： 对无线网的满意度

对于问题31 – 35，请你用下面的评分表达你对每种情况的满意度：

1 = 非常不满意 (VU)　　　4 = 满意 (S)
2 = 不满意 (U)　　　　　　5 = 非常满意 (VS)
3 = 一般 (N)

	情况	VU	U	N	S	VS
31.	总的来说，我对使用无线网完成任务的便捷性感到满意。	1	2	3	4	5
32.	无线网服务极大地影响了我搜索信息和管理学习的方式。	1	2	3	4	5
33.	我对信息技术部门提供的无线服务感到满意。	1	2	3	4	5
34.	总的来说，我对大学提供的无线网服务很满意。	1	2	3	4	5
35.	我很满意在校园里随时随地使用无线网。	1	2	3	4	5
36.	我很满意在宿舍可以使用无线网。	1	2	3	4	5

第五部分：人口统计学信息

请填写以下信息。

37.	性别：	❑ 男 ❑ 女
38.	年龄：	_____ 岁
39.	国籍：	_____
40.	在嘉兴学院学习的年数	_____ 年
41.	学习阶段：	❑ 本科生 ❑ 研究生 ❑ 博士生
42.	学院/中心：	_____
43.	院系：	_____

谢谢你的合作。

Index

Page numbers in *italic* indicate a figure and page numbers in **bold** indicate a table on the corresponding page

computer self-efficacy, 4, 5
  and lecturers' satisfaction with technology, 7, **7**, 10, *11*, 14, *15*, **17–18**, 21–22
  and students' satisfaction with online learning, 43–61, **49–50**, **52**, *53–54*, **54–59**
  and students' satisfaction with online research databases, 79–96, **84**, **87**, *89–90*
  and students' satisfaction with wireless internet, 107–126, **115–116**, *117–118*, **120**, *121–122*
COMSATS University Islamabad (CUI), 76, 77, 83
confirmatory factor analysis (CFA), 85, 88, 114
COVID-19 pandemic and online learning, 40–42, 46
CUI, *see* COMSATS University Islamabad (CUI)
culture and computer self-efficacy, 46

databases, *see* online research databases

ease of use, *see* perceived ease of use
educational inequality and online learning platforms, 42–43
EFA, *see* exploratory factor analysis
exploratory factor analysis, 2

four-facet re-specified measurement model, 14, 16, *17*
four-factor measurement model, 14, *15*, 85, 88, **89**

gender and students' satisfaction with wireless internet, 114, 119, **123**, 124

geographical differences and computer self-efficacy, 46, 53–55, **55–58**, 59, 60

HEC NDL, *see* Higher Education Commission National Digital Library (HEC NDL)
higher education
  assessing satisfaction with online learning in, 40–74
  assessing satisfaction with online research databases in, 75–96
  assessing satisfaction with wireless internet satisfaction in, 106–126
  assessing use of ICT in, 1–33
Higher Education Commission National Digital Library (HEC NDL), 76

ICT, *see* information and communication technologies (ICT)
information and communication technologies (ICT), 105
  and assessment of in higher education, 1–33
information literacy, 77
internet, wireless, 79–82, 92–93
  measuring satisfaction in higher education, 105–126
  questionnaire on measuring satisfaction, 131–137

learning management systems (LMSs), 40–41
lecturers
  and satisfaction with technology, 2, 4–20, **7**, **8–10**, *11*, **12–13**, *15*, 109
LMS, *see* learning management systems (LMSs)

ODAS model, *see* online database adoption and satisfaction (ODAS) model
online database adoption and satisfaction (ODAS) model, 6
online learning platforms
  background in China, 42–43
  questionnaire on testing satisfaction with, 68–74
  students' satisfaction with, 40, 44–74, **49–50**, *51*, **52**, *53–54*, **54–59**, 108–109
online research databases
  definition, 75–76
  ease of use and usefulness of, 6
  history of, 76–77
  questionnaire on assessing, 101–104
  students' satisfaction with, 76–96, **84**, *85–86*, **87**, **89–90**

perceived ease of use, 4, 5–7
  and lecturer's satisfaction with technology, **7**, 10, *11*, 14, *15*, **17–18**, 20–22
  and students' satisfaction with online learning, 43–74, **49–50**, *51*, **52**, *53–54*, **54–59**
  and students' satisfaction with online research databases, 79–96, **84**, *85–86*, **87**, **89–90**
  and students' satisfaction with wireless internet, 106–126, **115–116**, *117–118*, **120**, *121–122*, **123**
perceived usefulness, 4, 5–7
  and lecturers' satisfaction with technology, **7**, 10, *11*, 14, *15*, **17–18**, 21–22
  and students' satisfaction with online learning, 43–74, **49–50**, *51*, **52**, *53–54*, **54–59**
  and students' satisfaction with online research databases, 79–96, **84**, *85–86*, **87**, **89–90**
  and students' satisfaction with wireless internet, 106–126, **115–116**, *117–118*, **120**, *121–122*, **123**
PEU, *see* perceived ease of use

Rasch model, 1, 2
  and assessing satisfaction with online learning, 47, 48, **49–50**, *51*
  and assessing satisfaction with online research databases, 78, 84–85, *85–86*, 95

and assessing satisfaction with wireless internet, 111, 112, *113*, *114*
validating questionnaire on perceived ease of use and usefulness of technology, 7–10, **9–10**, *11*
regional differences and computer self-efficacy, 46, 53–55, **55–58**, 59, 60

satisfaction
  of lecturers with technology, 2, 4–20, **7**, **8–10**, *11*, **12–13**, *15*, 28–39, 109
  of students with online learning platforms, 40, 44–74, **49–50**, *51*, **52**, *53–54*, **54–59**
  of students with online research databases, 78–96, **84**, *85–86*, **87**, **89–90**
  of students with wireless internet, 106–126, **115–116**, *117–118*, **120**, *121–122*, **123**
self-efficacy, computer, *see* computer self-efficacy
self-esteem, 5, 21
SEM, *see* structural equation modeling (SEM)
social cognitive theory, 4, 44, 78
social presence, 5, 21
structural equation modeling (SEM), 1, 2, 16, 22, 111, 113–114
students
  satisfaction with online learning platforms, 40, 44–74, **49–50**, *51*, **52**, *53–54*, **54–59**
  satisfaction with online research databases, 78–96, **84**, *85–86*, **87**, **89–90**
  satisfaction with technology, 4
  satisfaction with wireless internet, 106–126, *113–114*, **115–116**, *117–118*, **120**, *121–122*, **123**

TAG model, *see* technology adoption and gratification (TAG) model
TAM, *see* technology acceptance model (TAM)
teachers, *see* lecturers
technology acceptance model (TAM), 3, *3*, 5–6
  and lecturers' satisfaction with technology, 21
  and students' satisfaction with online learning, 43–44, *44*
  and students' satisfaction with online research databases, 78–79
  and students' satisfaction with wireless internet, 106, 109

140 *Index*

technology adoption and gratification (TAG) model, 108, 110, 116
technology and lecturers' satisfaction with, 2, 4–20, **7**, **8–10**, *11*, **12–13**
questionnaire on, 28–39
technology satisfaction model (TSM)
hypotheses of, 5–7
and lecturers' satisfaction with technology, 2–33, *4*, *19*, **20**
and students' satisfaction with online learning, 41, 43, 47, **47**, 48, 50, **52**, *53–54*, **54–57**, 60–62, 107
and students' satisfaction with online research databases, 78–96, *80*, *85–86*, **87**, **89–90**, *91*, **92**, 107

and students' satisfaction with wireless internet, 105–126, *107*, *113–114*, **115–116**, *117–118*, **120**, *121–122*, **123**
theory of reasoned action (TRA), 3, *3*, 43, 78
TRA, *see* theory of reasoned action (TRA)
TSM., *see* technology satisfaction model (TSM)

unified theory of acceptance and use of technology (UTAUT), 81
usefulness, *see* perceived usefulness
UTAUT, *see* unified theory of acceptance and use of technology (UTAUT)

wireless internet, *see* internet, wireless